Universal Consciousness

A Scientific Exploration

Alexander Escobar, Ph.D

DEDICATION

To all my friends and family that have supported me throughout the years. Especially my wife Malia whose love is my rock and to my mother Alicia, who filled me with an enduring appreciation and love for education.

CONTENTS

ACKNOWLEDGMENTS

I thank all of those that have shared their feedback with me, including participants from several courses I taught based on this material. Chief among these are the members of Saint Bartholomew's Episcopal Church in Atlanta Georgia who attended a course on material from this book.

INTRODUCTION

Imagine a universe that pulses with awareness, vibrant and alive, unlike the cold and mechanical version that science often presents. This is the realm of Universal Consciousness, where everything, including our own human consciousness, is part of a grand interconnected tapestry. It's a profound concept that challenges the conventional boundaries of our understanding, stretching the limits of what we perceive as possible.

Through my forty years of studying and learning about nature through science, I feel that I have acquired an intuitive sense of natural phenomena. Consciousness is perfectly natural and the mistake we make is that we assume that what happens in our craniums is somehow different than what occurs in the rest of the universe. However, we know that the same rules of science that describe the entire universe also apply to what happens in our brains and vice-versa. This understanding leads directly to the question, "Could a fundamental form of consciousness exist throughout nature and the universe?"

Despite our scientific advancements, consciousness remains an enigmatic puzzle. Our own biases and reluctance to relinquish sole ownership of this phenomenon have hindered our quest for a deeper understanding. In doing so, we've unwittingly shut ourselves off from an all-encompassing wisdom, impeding our intellectual and spiritual growth as a species.

Consider the colossal challenges we face today, from the looming specter of global warming to the devastating decline of entire ecosystems. Merely relying on technological quick fixes or isolated groups of people to solve the problem won't lead us to solutions.

What we truly need is a fundamental shift in our perception—an awakening to the intricate connections between ourselves, the natural world, and Universal Consciousness.

Embracing Universal Consciousness has a profoundly personal importance as well. It's about recognizing that you are an integral part of this cosmic tapestry, woven into the very essence of the universe. Disregarding this truth is akin to disregarding a part of your own body like your shoulder, bumping blindly into doorways, and experiencing constant pain. You may wonder why life feels so agonizing, why a sense of loneliness and isolation pervades our modern societies. The truth is you are never alone. You possess a grandeur and untapped potential that surpasses your wildest imaginings.

To tap into Universal Consciousness, we must silence the incessant chatter that fills our minds, aggravated by our constant connection to the online world. It's time to shift our focus inward, reconnecting with our "innernet." Disconnect from the digital noise, find a quiet space, and attune yourself to the rhythm of your breath, the steady beat of your heart. When distractions arise, refocus. Joining others in this meditative practice can be immensely helpful, especially in the beginning. Seek out meditation groups and find the one that resonates with you.

Meditation serves as a conduit to clear away the mental clutter and perceive the subtle threads that bind us to the universe. You might explore different approaches, such as immersing yourself in nature or taking tranquil walks where you simply exist in the moment. It may take some trial and error, but the rewards are immeasurable.

While this book will delve into scientific intricacies, it's essential to begin with a profound truth. You are familiar with your five senses, but I believe there is also an inner sense. The way we humans sense true, heartfelt connection is through the emotion of love. Love is a sense and when I talk about connection to Universal Consciousness, I am also talking about a Universal Love.

Universal Consciousness, in its boundless reach, unifies us all. It offers the opportunity to tune into a sublime love that thrives within its depths. Throughout history, various religious traditions have developed techniques to attain this state of blissful connection. Prayer, as practiced by Christian mystics, is a powerful example. And now, this book presents yet another chance to access the universal bliss that awaits within Universal Consciousness—the birthright of every soul yearning for true connection.

1

UNRAVELING THE ENIGMA OF CONSCIOUSNESS

Prepare yourself for an intellectual adventure as we embark on a journey through the vast realms of consciousness. But before we delve into mind-bending concepts, let's start with the basics. The fundamental question that has intrigued thinkers throughout history, including countless curious souls like yourself, is: "What is consciousness?" While science has only begun to scratch the surface of this profound inquiry in the past few decades, I'll share some fascinating ideas that will stimulate your mind and propel us towards a deeper understanding.

When it comes to discussing consciousness, people often talk past each other, comparing apples to oranges. No wonder confusion reigns, and progress remains elusive. At its core, consciousness involves the intricate interplay of retrieving information from our environment, processing it, and responding accordingly. Picture yourself strolling through a tranquil forest, the crispness of the air kissing your skin. You instinctively reach for your coat, continuing your walk while marveling at the breathtaking tapestry of autumn foliage.

In more technical terms, we can dissect this process into sensation and perception. Sensation refers to the mechanical processes by which we absorb information about the world. Take, for instance, an odor molecule binding to a nasal receptor, triggering neural signals that traverse to the brain. Perception, on the other hand, occurs within the brain as it processes the incoming information. This neural activity magically transforms into a recognizable scent, creating an internal, subjective experience that may elicit pleasure or even prompt a desire to retreat.

It's crucial to recognize that this intricate dance of stimulus and response may occur without perception. At its simplest level, the interaction is purely mechanical. Imagine placing a speck of sugar in a solution teeming with *E. coli* bacteria. The microorganisms swiftly migrate towards the sugar, a phenomenon known as chemotaxis.

Observing this under a microscope might mistakenly lead you to believe that the bacteria are sensing and perceiving the sugar. However, chemotaxis is purely mechanical, a series of movements driven by the concentration gradient of the sugar, aiding the cells in obtaining energy.

A similar principle applies to the knee-jerk response in humans. How many times has a doctor struck your knee with a rubber hammer to test your reflexes? This procedure triggers stretch receptors in your muscles, sending signals to the spinal cord. Fixed neural circuits process this information, causing the relaxation and contraction of leg muscles. While other neural pathways convey the feeling of the hammer strike and your subsequent kicking leg to your brain, no conscious perception is required for this mechanical phenomenon to occur. It represents a slightly more advanced version of the responses seen in bacteria, thanks to the involvement of a nervous system, yet it remains strictly mechanical.

Moving to the next level, we encounter primary consciousness. At this stage, the stimulus-response dynamic takes on a different form, intertwined with inner subjective experiences of the outer world. Many animals likely possess this level of consciousness. Consider a deer in a meadow, attuned to its surroundings. When the sound of a snapping stick pierces the air, the deer's senses sharpen, anticipating the presence of a predator. Here, the stimulus is the air pressure waves generated by the breaking stick, causing fluid movements in the deer's inner ear and activating its auditory system. Neural signals transmit this information, leading to an internal subjective experience—a "snap" perceived by the deer—potentially triggering its fight-or-flight response. At this level, perceptions become real and play a pivotal role in processing the stimulus.

Our human consciousness aligns most closely with secondary consciousness. Here, the inner subjective experiences of primary consciousness intertwine with a sense of continuity through time, extending beyond the present moment. This ability allows us to develop a sense of self, detached from immediate circumstances. Such consciousness is believed to exist in a select few animals, including humans, chimpanzees, dolphins, and elephants.

As you can see, the distinction between stimulus-response scenarios carries significant implications when discussing consciousness. While it's easy to understand how machines and computers, as currently designed, can sense and respond to their environment in a mechanical manner, they lack the capacity for genuine perception. They lack the internal subjective experiences that many animals possess. This can be very confusing these days with people talking about artificial intelligence and other terms that mislead you into thinking that these machines somehow perceive their world. It is simply not so.

In this book, we will primarily focus on the rich mélange of human consciousness—the realm of conscious experiences we intuitively grasp in our everyday lives. Think of the sensations of vibrant colors, the symphony of sound frequencies we hear, and the myriad of tactile experiences, from gentle caresses to sharp pricks on our skin.

Even more, our conscious world encompasses the breadth of emotions that help us interpret the world around us. For example, love is one of our deepest human emotions. It possesses the remarkable power to heal, unite, and vanquish fear. Love is not only an integral part of our humanity but also a profound way of sensing our world. It emerges when we experience deep and heartfelt connections—a sensation that shifts our focus from ourselves to the intricate web of relationships we share with others and the environment.

Our inner conscious world is a complex tapestry, woven with sensory perceptions, emotions, and thoughts. This very complexity initially drew me to the field. However, when I expressed my interest in consciousness studies to my post-doctoral advisor in the 1990s, I was met with discouragement. Sadly, this sentiment was all too common in the natural sciences at the turn of the century. It took the groundbreaking work of Nobel Prize-winning biologist Francis Crick to finally convince others that the study of consciousness is a legitimate scientific pursuit.

I've often pondered why it took so long for the natural sciences to embrace the notion that consciousness can be studied scientifically. I believe that this hesitation is partly rooted in an intuition shared by my fellow scientists—a realization that a true scientific understanding of consciousness necessitates a seismic shift in our fundamental understanding of the universe and how it operates. To truly comprehend consciousness, we must embrace a paradigm shift even grander than Einstein's theories of special and general relativity. The magnitude of this change can be daunting and even terrifying to some.

As you continue reading, prepare to be astounded. Science will take us on a mind-bending expedition beyond the realms of your imagination.

2

QUALIA AND THE HIDDEN SYMPHONY OF AWARENESS

As I sit here writing this book, the vibrant colors of fall surround me, painting a breathtaking display at every turn. The ginkgo trees, with their brilliant yellow leaves, appear like ethereal clouds drifting along the streets. The white oak trees, adorned with rust-colored foliage, seem to wave at me as I pass by. Taking a stroll in the park has transformed into a dazzling kaleidoscope of hues, almost disorienting in their stunning vibrancy. In the midst of this captivating natural beauty, a question arises within me: How am I able to experience all these wonders? How am I consciously perceiving these mesmerizing colors and shapes?

So, let's delve into the enigma of conscious experience. While many ideas have been proposed, no current hypothesis provides a fully satisfying answer. It's time to explore consciousness from a different perspective—a biological standpoint. As a biologist, I naturally gravitate towards understanding consciousness and awareness through the lens of biology. Specifically, I'm intrigued by the primate striate cortex and its meticulously organized neural columns. In biology, we understand that structure leads directly to function. The Quantized Visual Awareness* (QVA) hypothesis predicts that this intricate structure of the striate cortex leads directly to the production of conscious visual experiences.

Within the striate cortex, there are circuits known as microcolumns, which some propose to function as independent processing units. My belief is that these distinct circuits generate individual fragments of

---

* I have published several peer-reviewed journal articles on the QVA hypothesis over the past decade (see About the Author at the end of this book). I will not delve deeply into the technical aspects of this hypothesis within this book. Instead, I aim to focus on the philosophical implications of these ideas, as I believe they hold significant implications for our understanding of ourselves and the natural world.

visual consciousness within the striate cortex, which later coalesce into the comprehensive visual experiences we humans enjoy. While my research primarily focuses on visual experience, I believe these principles apply to all our senses. I begin with vision because it has been extensively studied, but I want to emphasize that every sense is equally valuable and important. With these philosophical implications in mind, I invite you to explore these concepts further.

In the past, scientific study of our personal experience of the external world seemed implausible. Our subjective experience appeared to be fundamentally distinct from the rest of nature, rendering it beyond the realm of scientific investigation. Central to this belief was the notion that our conscious experience couldn't be exhibited or objectively studied by others. However, contrary to these propositions, there's no reason to assume that our subjective experience lacks a physical component or that it cannot be scrutinized through the lens of science.

Indeed, many neuroscientists now contend that our inner subjective experiences, such as vision, are products of the brain's physical activity. A wealth of evidence supports this perspective, including studies of patients with damaged visual cortices who lose specific visual abilities like color perception (achromatopsia). Within the fields of neuroscience and cognitive science, there is a consensus that neural activity within our brains gives rise to our conscious experiences.

In the realms of psychology and philosophy of mind, scholars have contemplated the notion of qualia. Qualia represent the smaller aspects of our subjective experience, such as the patch of blue sky we perceive on a cloudy day. By considering our sensory experiences in this way, we can posit that our overall experience emerges from the sum of these sensory qualia.

While the concept of qualia remains a topic of debate for many scholars, I firmly believe in their existence. Qualia are the fundamental units of conscious experience, akin to how the rest of the universe seems to be comprised of fundamental units depending on the domain of study (atoms, cells, photons). Though we have artificially separated our subjective experience from the objective world, I believe it to be a misconception. Our conscious experience operates in a manner analogous to the rest of the universe, with basic units of subjective experience possessing physical attributes.

Depending on your perspective, these assertions about conscious experience may appear radical or mundane. What I propose is that conscious experience is entirely natural—it's an integral part of nature. Consequently, we can explore and study it scientifically, just as we do with other phenomena in the universe. Moreover, conscious experience possesses physical properties. For reasons I will elaborate on later, I hypothesize that the physical counterparts of qualia are the electromagnetic fields (EMFs) generated by neural circuits with distinctive structures. Like other foundational elements of nature, I believe these units of awareness are distributed throughout the natural world, primarily residing in the nervous systems of the countless animals that populate our world.

The hierarchy of structure is a basic principle of biology. The concept is based on the idea that smaller subunits come together to form more complex wholes. This applies to all levels of organisms since molecules form cells, cells form tissues, tissues contribute to organs and so on.

Evolution has selected for all biological systems to work in the same manner. Why should we believe that conscious experience is an exception to the rule?

Renowned philosopher of mind David Chalmers has posited that consciousness should be regarded as a fundamental aspect of nature. I wholeheartedly embrace this notion and further propose that evolution has ingeniously organized this fundamental aspect into selectable units known as qualia. While these qualia are simple and small individually, their cumulative effect gives rise to significant experiences, such as our complex human visual field.

I am not alone in these assertions. Other thinkers have also postulated that smaller forms of consciousness can be generated within our brains. The distinguished neurobiologist Semir Zeki, for instance, introduced the concept of micro-consciousnesses over a decade ago, associating specific, independent types of visual consciousness with the activity of visual centers or circuits. Mounting evidence suggests that visual consciousness manifests as qualia arising from the action of individual microcolumns (Figure 2.1) within the striate cortex of primate brains. While they may exist in other locations as well, they are abundantly present in the striate cortex, making it an ideal starting point for exploration.

In the context of consciousness, individual qualia correspond to the simple attributes of our sensory experiences. With vision, we could say there are qualia responsible for the sensations of color, depth, motion, and line orientation at specific points in our visual field.

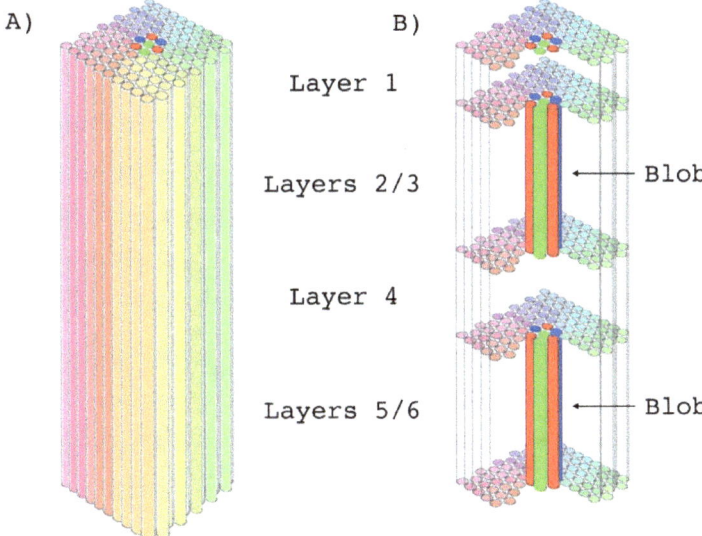

Figure 2.1 50-80 microcolumns are bundled into one ocular dominance column - ODC (A). Each ODC corresponds to one point of our visual field. The pastel colored microcolumns create line orientation qualia and the red, blue, and green microcolumns in the middle of the ODC create color qualia (B). Blobs are tissue sensitive to colored stimuli. This architecture allows for the exquisite control of the visual experience for each point of our visual field.

The types of qualia are likely limited. For instance, there may only be three color qualia—red, blue, and green—that combine to create the full spectrum of colors. The contribution of any one quale to our overall experience is incredibly minute.

To illustrate this idea, imagine a computer screen. Despite its ability to display an extensive array of scenes, we know that all of them are constructed from just three types of pixels—red, blue, and green. While it may be challenging to perceive an individual pixel, the collective presence of these pixels generates the rich variety of images we see on computer displays.

Neural microcircuits with specific structures reproducibly create the same type of quantized awareness. In other words, there is a specific circuit shape that creates the sensation of red whenever it is activated. This bit of awareness is the same no matter where it is produced. Thus, a circuit with the same structure will yield the same quantized bit of red for you and for me. In this way, you can think of these quanta of consciousness as elemental atoms. A specific atom should be the same and have the same physical properties whether it is in your hand or somewhere else on the planet. Moreover, it is probable that just like a limited set of atoms can create all possible structural forms, we can create all possible experiences from a limited set of qualia.

Qualia have been harnessed by evolution to enhance the survival of the organisms harboring them. At some point in evolutionary history, a neural circuit produced a bit of awareness due to its structure. If this conferred a selective advantage, there may have been a subsequent proliferation of such awareness. In biology, we refer to this process as exaptation—a trait being selected for a purpose different from its original function. Importantly, it's worth considering that these early bits of awareness might have been employed differently than how we utilize them today.

As time passed, the number and variety of qualia expanded in species that possessed them. Additionally, the complexity of how these qualia were organized increased. It's highly probable that qualia emerged early in the evolutionary trajectory of animals and have since permeated the animal kingdom. While one could argue that animals with eyes don't genuinely perceive anything visually, the correlation between eyes and visual perception is readily apparent. Given that so many organisms possess some form of visual apparatus, it's likely that numerous species also generate visual qualia.

Our perception is the sum of all these bound bits of awareness, much like our bodies are the sum of the cells that contribute to our physical form. This perspective takes a bottom-up approach, wherein perception arises from the integration of individual bits of awareness into a coherent experience. From this vantage point, we realize that we are not the architects of our conscious experience but rather the beneficiaries of all these quanta of consciousness working in unison.

Consequently, the number, type, and organization of qualia involved largely define our subjective experience. By understanding subjective experience in this manner, we establish a foundation for its scientific investigation through established principles.

But how does it all come together? When we contemplate the integration of these separate fragments of subjective experience into one overarching experience, questions inevitably arise. It becomes even more perplexing when we consider that these independent neural circuits are physically dispersed within our craniums. If we only view neurons and neural circuits as information processors, as is common in neurobiology, it becomes difficult to envision how this integration occurs.

The key to understanding how all these independent qualia are bound together into one conscious experience comes from realizing that the physical surrogates of qualia are not the neural circuits themselves, but the specifically shaped electromagnetic fields (EMFs) produced by these neural circuits. The electric currents moving through microcolumns create EMFs with unique topologies and these automatically integrate themselves into the brain's larger EMF. In this way, any part of the brain can contribute instantaneously to our overall conscious experience and generate the rapid changes we experience in our everyday consciousness.

Now, let's address the "Hard Problem," as coined by David Chalmers. The challenge lies in explaining how neural excitation produces a conscious experience when we can have neural excitation without consciousness—such as in the knee-jerk reflex, which occurs unconsciously. Why should specifically shaped EMFs generated by certain microcolumn circuits possess any associated awareness or consciousness?

The answer lies in recognizing that biological systems do not create universal physical properties from scratch; instead, they exploit existing properties of nature to enhance their reproductive success. For example, multicellular organisms didn't invent diffusion but evolved to utilize it in distributing molecular oxygen throughout their bodies. Biological systems harnessed the power of chemical reactions for daily metabolism without inventing these reactions themselves. Electromagnetism existed in nature long before biological systems emerged, and yet many organisms, like electric eels, have evolved to exploit electromagnetic phenomena to enhance their survival.

If we consider that biological systems generally don't invent universal physical properties, we can reasonably ask whether some fundamental form of awareness already exists in nature. Could it be that a subtle form of conscious awareness is embedded within all electromagnetic fields in nature? In such a scenario, biological systems have evolved to leverage this subtle form of electromagnetic field awareness, generating simple yet specific types of awareness within our brains.

Electromagnetic fields are wave mechanical phenomena, subject to the principles of constructive and destructive interference. We witness this when water wave troughs and crests intersect, leading to either cancellation (destructive interference) or amplification (constructive interference). In inanimate objects, electromagnetic fields of atoms and molecules interfere randomly, resulting in an overall reduction (destructive interference) of integrated fields. As a result, inanimate objects typically don't produce significant electric or magnetic fields at a macroscopic level. As such, you would expect any subtle form of consciousness associated with these fields to be minimal.

The QVA hypothesis predicts that cortical microcolumns precisely shape EM fields through the constructive and destructive interference arising from various parts of the microcolumn. These microcolumn EM fields correspond to simple forms of conscious experience (qualia) like a bit of red or blue. In our brains, millions of these qualia EMFs bind together automatically as they are produced and create the complex visual field we experience every day.

For over ten years, I had the privilege of participating in the Emory Tibet Science Initiative (ETSI), which aimed to integrate modern science into the monastic curriculum of Tibetan Buddhist monks in India. During my visits to various monasteries, I engaged in discussions with Geshe monks (akin to Ph.D. scholars in our system) about diverse topics. In one such conversation, I presented the ideas of consciousness I've shared with you here. To my surprise, I noticed the monks smiling at me.

Curiosity piqued; I asked the Geshes how they believed consciousness arises. They proceeded to describe something strikingly similar to the concepts I had presented from a Western scientific perspective. In fact, at one point, a Geshe remarked, "It's like what you described." Through their own empirical methods of study, these monks have detected a subtle form of awareness permeating nature.

This convergence of ideas could be dismissed as a mere coincidence, but it also exemplifies the way science should ideally function. Scientific inquiry should lead us to similar conclusions regardless of our starting point. The notion of a subtle form of awareness existing within electromagnetic fields may be the most profound insight emerging from this line of research. Take a moment to contemplate the implications if conscious awareness is indeed embedded within the electromagnetic fields that permeate nature. Does your relationship with nature undergo a transformation? Considering that you are an integral part of nature, can you truly separate the conscious awareness arising in your brain from the subtle form of Universal Consciousness that may surround you? How might our understanding of the universe change if we were to accept its inherent consciousness?

3

JOURNEYING INTO THE CURIOUS REALM OF SPECIAL RELATIVITY

You can think of an electromagnetic field (EMF) as a patterned field of photons since photons carry the electromagnetic force (Fig. 3.1). EMFs have a lot to do with photons and electromagnetic radiation. If we truly want to understand the conscious electromagnetic fields we discussed in the last chapter, we need to explore photons and EM radiation more fully. So buckle up, fellow cosmic explorers, as we embark on a journey to unravel the mysteries of photons and EM radiation. It's a voyage of discovery that promises to shed light on the very essence of consciousness itself. Who knows what secrets we'll uncover along the way?

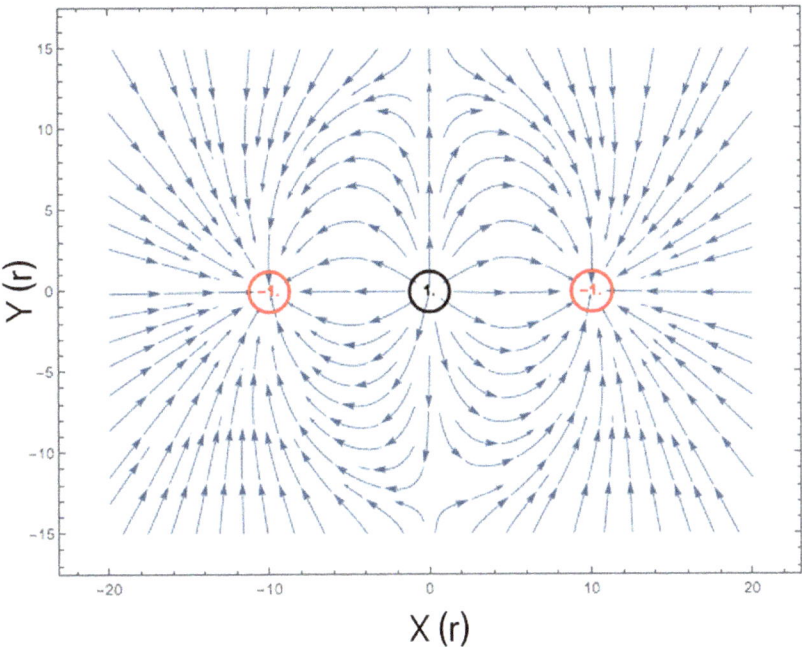

Figure 3.1 The image shows an electromagnetic field created by three point-charges. The field lines show the structure of the electromagnetic field and represent the electromagnetic force between these point charges. The electromagnetic force is carried by photons- visualize the arrow heads shown above as photons.

Imagine yourself as a photon, a tiny particle of light, soaring through space at unimaginable speeds. Photons, like little marbles, zip through the cosmos, travelling from one end of the universe to the other.

But here's the twist: photons aren't just particles, they're also waves. Just like the mesmerizing waves crashing on the shore, light waves have their own unique patterns of troughs and crests. It's mind-boggling to think that something can be both a particle and a wave, but that's the enigmatic nature of light.

These wave-like photons form what we call electromagnetic radiation, which spans a vast spectrum of frequencies (Figure 3.2). We humans can only perceive a small sliver of this spectrum as visible light. So, when we talk about light, we're referring to this limited range of electromagnetic radiation that we can see with our eyes.

The Visible Spectrum

Visible light ranges from 380 nm to 760 nm in wavelength.

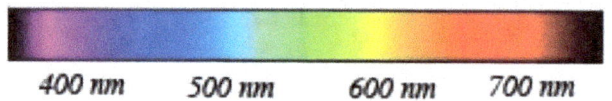

400 nm   500 nm   600 nm   700 nm

m - meter (approximately one yard)
nm - nanometer (one billionth of a meter)

Figure 3.2 The light we see falls into a range of electromagnetic wavelengths that range from approximately 380-760 nanometers.

Let us pause for a moment and consider the sheer speed at which photons travel. Picture this: it takes a mere 8.3 minutes for a photon to journey from the scorching surface of the sun to the Earth, spanning a staggering distance of 147 million kilometers. That's because photons dash through space at a mind-boggling speed of 300,000 kilometers per second. When light bounces off surfaces and enters our eyes, it's captured by our retinas, enabling us to see the world in all its splendor.

Albert Einstein radically changed our notions of space and time. Before Einstein came on the scene, the world was stuck in the classical Newtonian worldview. This worldview is more or less based on our common modern experience. For example, time and space were considered fixed, unchangeable characteristics of the universe. Although this worldview worked well for most things, there was one aspect of reality that did not agree with this understanding. This strange phenomenon dealt with the speed of light. The scientific breakthrough that came from studying this puzzle is Einstein's theory of special relativity.

We know the speed of light is 300,000 kilometers per second. This is true if one is standing on Earth or on some far distant planet in another galaxy. This may not seem odd to you. After all, light is light, and it makes sense that it should behave the same in different parts of the universe. However, there is a strange twist to all of this. Not only is the speed of light constant from place to place, it is also constant from speed to speed.

Imagine standing on the side of a road, measuring the speed of passing cars with a radar gun. For the sake of argument, let us say that you have a powerful measuring device that can accurately tell you the speed of any moving object relative to you. As you measure the speed of the passing cars, you find they are flying by at 100 kilometers per hour (kPH). If you were to get into your car and pursue one of these vehicles at 95 kPH, your measuring device would let you know that the car ahead of you is moving at 5 kPH relative to you.

Now if you were chasing after light-speed photons at close to the speed of light, how quickly would you expect the photons to move relative to you? If you apply the logic from above, you would expect the photons to slowly accelerate away from you. After all, you are almost going the same speed. However, light once again acts in an unpredictable way. If we were to jump into an ultra-rocket and chase after a beam of light at close to light speed, your measuring device would tell you that the photons were moving away from you at light speed and not just a little faster like the car example above. This is true no matter how close we get to light speed. It defies our intuition but getting past our intuition was part of Einstein's genius.

The speed of light is constant no matter how fast you are moving. If you are moving toward a light beam or away from light, the photons will always appear to travel at the speed of light. This hypothesis has been tested repeatedly, and it always holds true. Nobody knows why light always appears to move at light speed, but we do know some strange effects that arise from this phenomenon.

For example, an astronaut in a rocket speeding at close to light speed will perceive everything as normal except that no headway can be made against the beam of light. To someone standing outside of the rocket, things will appear quite differently. If this external observer were to look through the porthole of the passing spaceship, they would see the speeding astronaut moving very slowly within the ship, as if the astronaut were moving in slow motion. Also, the astronaut would seem to have shrunk through the dimension that the rocket is moving (Figure 3.3).

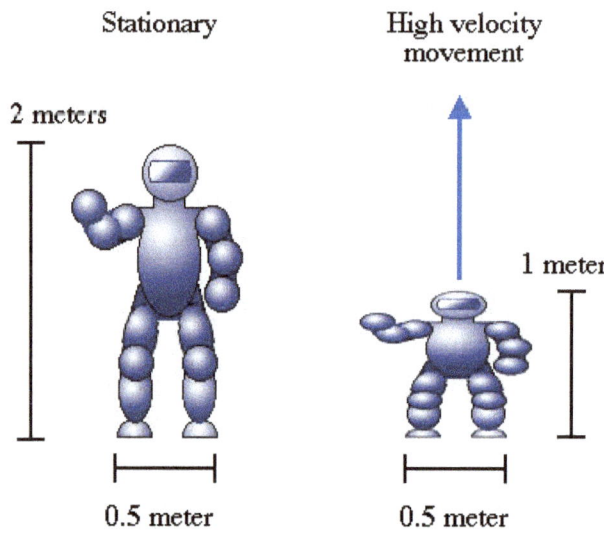

Stationary

High velocity movement

2 meters

1 meter

0.5 meter

0.5 meter

Figure 3.3 At velocities that approach the speed of light, objects shrink along the dimension that corresponds to their movement. In the figure above, we can see that our space traveler has decreased to half their original height but retains the same width and thickness.

The faster the rocket went, the slower and smaller our space voyager would seem. You may be thinking these strange observations are some sort of illusion due to speed, but this is not the reason.

What has happened is that unchangeable time and space have done just that—changed! Time is actually passing more slowly for the astronaut, and all things that occur within the spaceship occur at a slower pace. This is called time dilation. The wristwatch of our speeding astronaut is recording time at a slower rate than the watch of the observer on the outside. The faster the astronaut goes, the slower the hands of the astronaut's wristwatch will move. The astronaut will also continue to shrink along the dimension of space that corresponds to the movement of the ship. This is called Lorenzian contraction. *

---

*For those of you interested in the mathematics, Einstein's equation used to describe the effects on time is: $\Delta t = \Delta t_0 / (1 - v^2/c^2)^{1/2}$, where Δt_0 is the time period observed by the fast moving astronaut, and Δt is the elapsed time for the stationary observer. Notice that as v (the velocity of the fast-moving astronaut) gets close to c (the speed of light), the denominator of this fraction becomes very small. When you divide any number by a very small number, you get a very large value. In other words, Δt becomes very large as the astronaut gets close to the speed of light, and the stationary observer sees the time period for the moving astronaut become very large. Another way of saying the same thing is that the stationary observer witnesses the astronaut aging much more slowly since each instant that passes for the astronaut will seem like years or much longer for the stationary observer.

A similar analysis gives results that are just as interesting for the property of space. The effect of high velocities on space is given by the equation: $L = L_o(1 - v^2/c^2)^{1/2}$, where the proper length L_o is measured by the speeding astronaut and L is the length measured by the stationary observer. Again, note that L approaches zero as v gets close to c, indicating that the astronaut is becoming flat in the dimension of travel.

What does this say about time and space? If our astronaut were speeding at very close to light speed, their 2 meters (six feet) at some point would shrink down to 1 meter (3 feet), yet they would still perceive themselves as standing 2 meters tall (Figure 3.4). Also, time is passing more slowly for our astronaut, yet to them it would seem to be passing at a normal rate. Could it be that the way we perceive space and time is a product of the mind? Is the way we understand time a product of our form of consciousness or shaped by our perception?

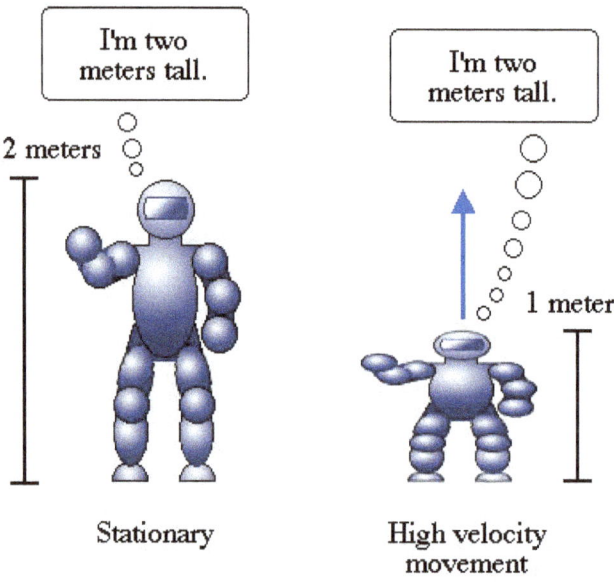

Figure 3.4 The high velocity figure has shrunk due to the cosmic voyager's high speed but still perceives themself and all other things within the ship to be the same height.

Our conventional notions of fixed time and space crumble under the weight of special relativity. Time becomes malleable, passing at different rates for different observers. Space loses its fixed dimensions, warping and contracting as objects move at high velocities. The universe reveals its complexity, challenging our simple interpretations.

Time for a thought experiment! Imagine you are a photon and that you can vary your speed (there is no such thing, but a nice tool for our thought experiment). As you increase your velocity, the more you will appear to slow down and shrink along the space-dimension through which you are moving to an external observer. This will continue, until at the speed of light, what? Amazingly, at light speed, time will finally stop and the dimension through which you are moving will contract down to zero. In other words, you as the photon will have no thickness in the direction you are moving and will exist in a timeless state.

This may all seem strange, but special relativity predicts this outcome. Since the theory has held out against so much experimentation, there is no reason to doubt this prediction of special relativity. Despite its non-intuitive nature, the theory indicates that light exists outside of time. Photons do not age, and a photon that has been traveling through the universe since the beginning of time is as new today as the moment it was created. This must be true for all light and electromagnetic radiation. Thus, a significant part of our universe that includes all visible light and electromagnetic radiation exists in this same timeless and spaceless state.

There is a symmetry that occurs in the description of the photon above. As the photon moves through the universe at light speed, it is also valid to say the universe is moving past the photon at light speed in the opposite direction. This is similar to what you experience when you drive down a street and pass trees on the side of the road. From a physics standpoint, it is equivalent for you to say you are passing the trees at sixty kilometers per hour or to say the trees are passing you at the same speed in the opposite direction.

Because of this symmetry, we can assign the role of the external observer to the photon. As the photon looks out at the universe passing by it at light speed in the opposite direction, the same properties that applied to the speeding photon would also now apply to the universe. Namely, the universe the photon "perceives" is frozen in time, and the dimension through which it is moving past the photon has zero length. Since we are talking about the entire universe, you can see this is a significant statement. The entire dimension of the universe through which the photon is traveling would have zero length, and therefore the photon is traveling through no space!

A photon traveling from a distant part of the galaxy into your eyes would "feel" as if it moved from its starting point to its ending point instantaneously. This actually makes sense since the dimension of the universe through which the photon is traveling has zero length and it should take no time to travel zero distance. To the photon, it will appear as if it is here and there simultaneously. Light exists in a state of only "here and now."

This may seem paradoxical, but special relativity predicts this outcome. A photon traveling through the vast expanse of the universe seems to traverse distances at 300,000 kilometers per second. However, from the perspective of the photon, it is both here and there simultaneously, free from the constraints of time and space.

This revelation challenges our preconceived notions of time and space. There is no absolute frame of reference. The viewpoint of the photon is just as valid as that of the external observer. Time and space become fluid, varying with the observer's perspective.

The electromagnetic force is carried by photons. Photons are electromagnetic radiation. Since all electromagnetic radiation exists in a timeless, spaceless state, it follows that all electromagnetic fields exist in this same state. We have previously explored the idea that a subtle form of awareness is embedded in electromagnetic fields. We shape this subtle form of awareness into very specific instances of conscious experience within our brains (qualia). Therefore, our conscious experience must also exist in the same timeless, spaceless state we have discussed in this chapter. Both propositions are true. Our experience arises from the neural activity of the brain, and simultaneously, this experience exists in a timeless, spaceless state due to its electromagnetic nature.

This opens up a cascade of questions: Is it possible to have multiple timeless and spaceless states that are here and now or is there only one? If there is only one, is it possible that all consciousness comes together into this one state? What properties would we expect for such a massive form of consciousness? In the upcoming chapters, we will venture further into these enigmatic territories, seeking answers that push the boundaries of our understanding.

4

THE SINGULARITY OF LIGHT

Welcome to an extraordinary journey that transcends our ordinary understanding of space and time. In this chapter, we will delve into the profound ideas of two great thinkers: Albert Einstein and Richard Feynman. Their contributions to science revolutionized our understanding of nature and brought us to the edge of our intellectual boundaries. Brace yourself as we explore the wavelike nature of light and the mind-boggling concepts it entails.

Let's start by envisioning the mesmerizing dance of ocean waves. Picture yourself at a picturesque harbor, where an artificial barrier stands in the way of incoming waves. As the waves approach the coast in parallel lines, they abruptly halt upon reaching the barrier, except for the openings designed for boats. Something enchanting happens at these openings. The waves passing through them transform from straight and parallel lines into graceful curves, as if some aquatic giant dropped stones into the water, creating concentric ripples (Figure 4.1).

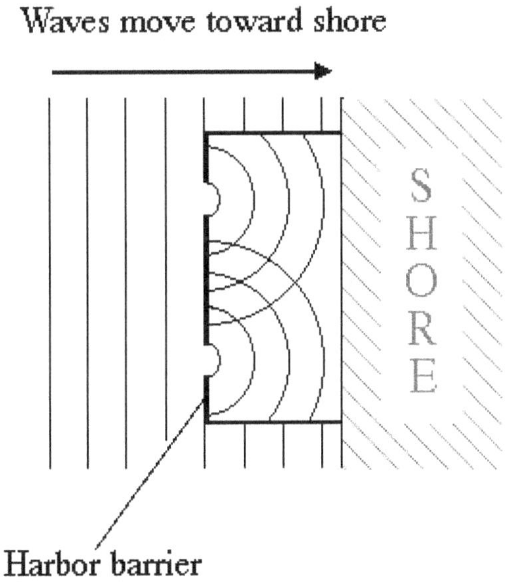

Waves move toward shore

Harbor barrier

Figure 4.1 The straight waves become concentric waves as they enter the harbor through the portals. This process is common to all wave-like phenomena.

When two openings are close together, the concentric waves produced by each portal interact in unique ways. When wave crests align, they amplify each other, resulting in even higher crests. Conversely, when troughs align, they deepen, creating more profound troughs. But what happens when a crest meets a trough? They cancel each other out, leaving a flat surface in their wake. These mesmerizing patterns of wave interaction are known as interference patterns and are observed in all wave phenomena (Figure 4.2). When we look at the back wall of our harbor and measure the height of the waves on the wall, we find the interference pattern clearly manifests itself on this back wall. There are points on the wall at which we always find large fluxes in the wave height and there are places that appear to be at a dead calm.

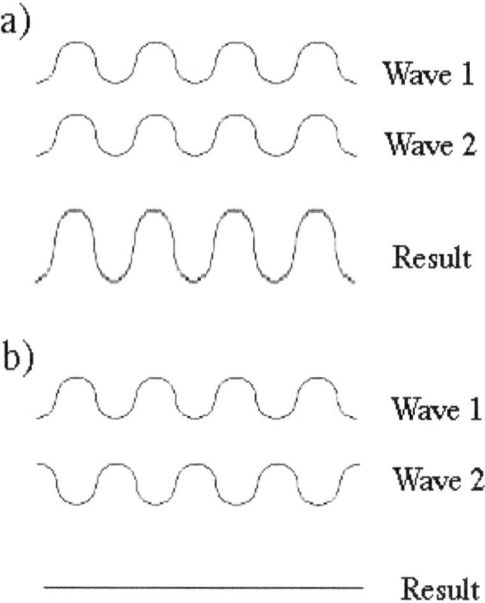

Figure 4.2 (a) When two waves line up such that their crests and troughs overlap, the result is a single wave with higher crests and deeper troughs. (b) When two waves are out of synch such that the troughs of one match with the crests of the other, the two waves cancel each other out and create a flat surface.

Now, let's shift our focus to light. Imagine shining a laser flashlight at an opaque board containing two razor-thin slits side by side. Behind the board, there is a wall onto which the light passing through the slits will project (Figure 4.3). What do you expect to see on the wall?

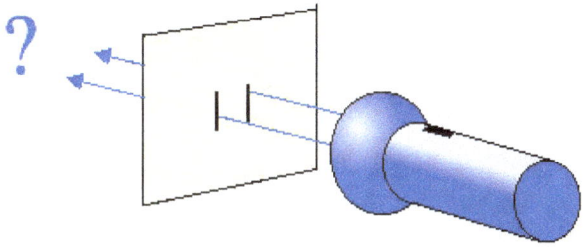

Figure 4.3 Light from our flashlight will pass through the two thin slits of the opaque board. What do you expect to see on the wall behind the board?

If light behaves as particles, we might anticipate two razor-thin lines of light on the wall, created as the particles shoot through the slits. However, if light behaves as waves, we might expect something more intriguing, reminiscent of the harbor's interference patterns. Did you guess it? Experiments, known as the double-slit experiment, reveal interference patterns when light passes through the two slits (Figure 4.4).

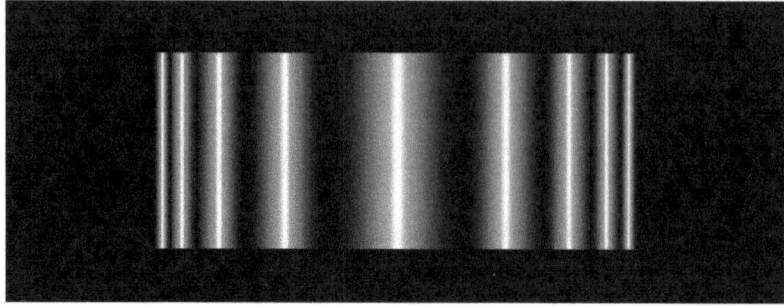

Figure 4.4 When the troughs or crests of the light waves add up, we get a region of brightness. When the troughs and crests cancel each other out we get a region of darkness. This is an interference pattern and is analogous to what we would expect to see on the wall at the back of our imaginary harbor.

Thomas Young, an English scientist, first conducted the double-slit experiment in the 19th century. Since then, it has been repeated in various ways, even using single photons passing through the slits one at a time. Surprisingly, despite releasing photons individually, the cumulative result on the screen behind the slits still produces an interference pattern.

Given that we already know that an interference pattern is produced when light passes through the two slits, the results of this slower experiment do not seem very exciting. It appears that we are repeating the same experiment at a slower pace. However, the results of the single-photon experiment are extremely shocking! Remember, in the single-photon experiment, photons are passing one at a time through either slit. How can an individual photon interfere with itself as it passes through one slit or the other? At the end of this experiment, we are somehow generating an interference pattern on the back screen, but this defies explanation.

This mind-boggling phenomenon has puzzled scientists for over a century. Richard Feynman, one of the greatest theoretical physicists of the 20th century, proposed an interpretation known as "sum-over-paths." According to Feynman, each photon simultaneously takes all possible paths through space and time to reach its final destination on the screen (Figure 4.5). By adding up the contributions of all these paths, we arrive at the predicted interference pattern. Feynman's ideas are highly respected by mainstream science, and he is given as much credibility as Einstein and for good reason. His work led to the development of the field in physics known as quantum electrodynamics. Feynman was a genius along the lines of Einstein; he could see past common sense and get to a place that transcended our everyday understanding of reality.

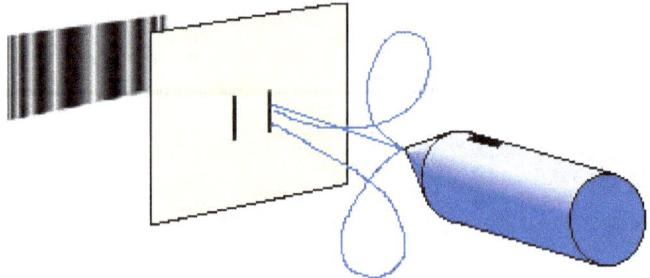

Figure 4.5 Each individual photon simultaneously travels all possible paths to reach its destination on the wall behind the screen. The lines that emanate from the light source represent some of these paths. Eventually, a diffraction pattern is generated on the wall even though photons were released one at a time.

But what does it mean for a photon to take all possible paths simultaneously? To understand this, we must revisit Einstein's theory of special relativity discussed in Chapter 3.

We know that light travels in all directions throughout the universe. According to Einstein's theory of Special Relativity, each photon should perceive itself as traveling instantaneously, with the dimension it traverses having zero length. Every photon exists in a timeless and spaceless state. Now, let's ponder a fascinating question: Do photons occupy different states of zero space and time in the universe or is there really only one state with these properties that includes them all?

Think about the number line. If we ask, "How many numbers have a quantity?", there are an infinite number of possibilities. However, if we ask, "How many points have no quantity?", there is only one answer. Only zero represents the absence of any quantity. Zero is the only way to have no quantity and if so, perhaps there is only one way to have zero space and time—one way to exist in a timeless and spaceless state.

To explore this question, let's play with a 3-D Cartesian model of space (Figure 4.6). In this model, each point represents a unique location in space described by its x y, and z coordinates. However, there is a special point known as the origin, where all three axes intersect, and all values are zero. This point holds a profound significance.

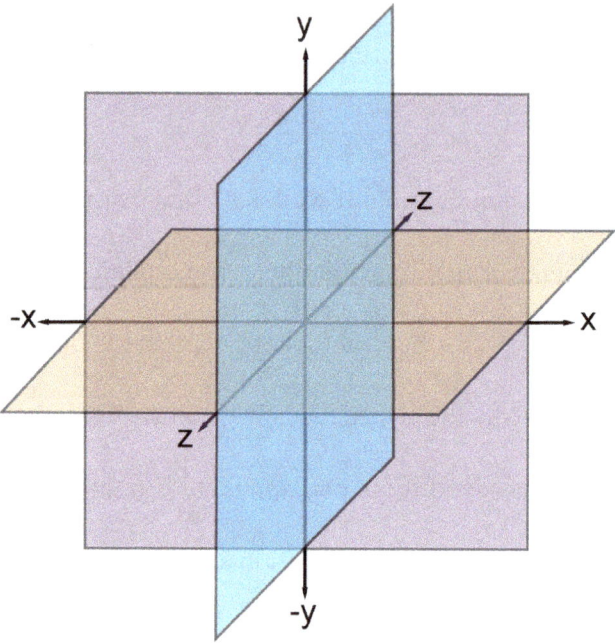

Figure 4.6 A 3-D Cartesian model is shown. The 2-D planes created by each set of two axes are shown in the figure to illustrate the 3-D nature of the model. All number lines extend out to infinity in either the positive or negative direction.

Imagine that our Cartesian space represents three-dimensional space, with x, y, and z dimensions and with the origin at the center of our universe. Now, envision light traveling through this space. From our perspective, the photon's path would trace out a line segment in the Cartesian space. But from the photon's perspective, no distance is traversed. In the photon's framework, all the points along its path collapse onto one another, merging into a single point. If we extend this concept to all the light in the universe, the entire Cartesian space folds and shrinks along every dimension, ultimately converging at the origin.

Let's employ another model to illustrate this concept. Envision our universe as a two-dimensional membrane, spherical in shape (Figure 4.7). As time progresses, the sphere representing our expanding universe grows larger. At the center of this space-time sphere lies a singularity, the origin of the universe, and the distance from this point to the expanding membrane represents the passage of time. In other words, the radius is a measure of time.

Now imagine that a photon from one side of the universe has traveled to the other side of the universe. From the perspective of the photon, this sphere has pinched itself such that all the points that describe its path are one and the same. We know that light travels from all parts of the universe to all other parts. If we were to repeat this step innumerable times (once for each photon in the universe) we would be left with a point and this point would be centered on the original singularity.

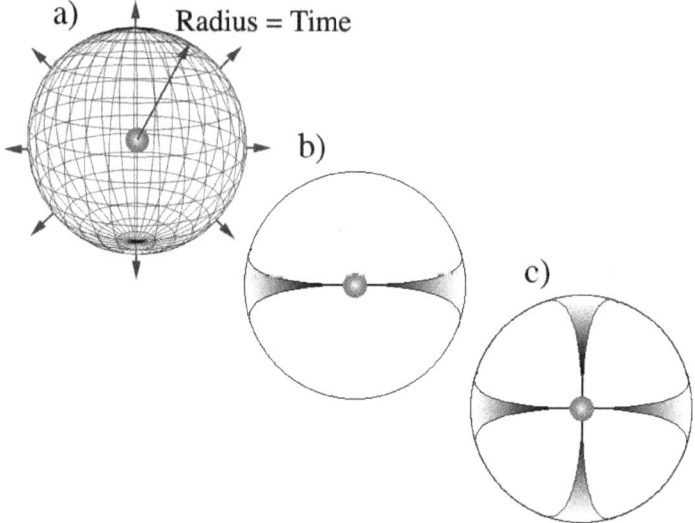

Figure 4.7 a) Our common three-space dimensions are shown here as a two-dimensional spherical membrane. This membrane is expanding over time, and thus, the radius of the sphere is equivalent to the elapsed time since the beginning. The beginning is shown as a point in the center of the sphere, and this corresponds to the singularity at the beginning of time and space. b) From the perspective of light, the membrane pinches or folds such that two distinct points in space-time come together. Light is both at its starting point and ending point simultaneously. c) If you repeat this step to represent all photons, the membrane will continue to pinch itself, and if this is done innumerable times, the membrane will pinch or fold itself down to a point that will be coincident with the singularity at the center of this model.

If we combine the "perspectives" of all of those traveling photons into one all-encompassing "perspective", all space-time shrinks down to a point. This point would include all electromagnetic radiation from the beginning of time and for the entire universe. This timeless singularity would include all pasts, presents, and futures simultaneously in an *eternal instant*. Since photons exist throughout time and space, the light of this and every moment contributes to this singularity.

If all space-time points traced out by every light path in the universe are included within this singularity, then every photon has access to all those space-time points. All photons sample every possible light path just by existing in this state and every photon should behave as if it is connected to all possible light paths simultaneously. Wait a minute, isn't this exactly what Feynman predicted with his sum-over-paths explanation?

In modern physics, this type of connection is known as quantum entanglement. The idea here is that all light exists in a state of quantum entanglement since all light coexists within the same timeless, spaceless state. The manifestation of this is that all photons have the property of "sum-over-paths". In other words, all photons behave as if they are taking all paths simultaneously in the double slit experiment due to their existence in this shared timeless and spaceless state. They are both a single photon and connected to all other photons through quantum entanglement at the same time.

Within science, the singularity at the beginning of the universe is thought to have had no dimensions, no time, and to be massively energetic (Figure 4.8). This source of all that exists contained the entire

universe within itself before it gave birth to the cosmos in the ultimate

act of creation known as the Big Bang. Let me propose to you that

these two singularities must be one and the same. Understanding this

concept requires that we readjust how we think about time and space

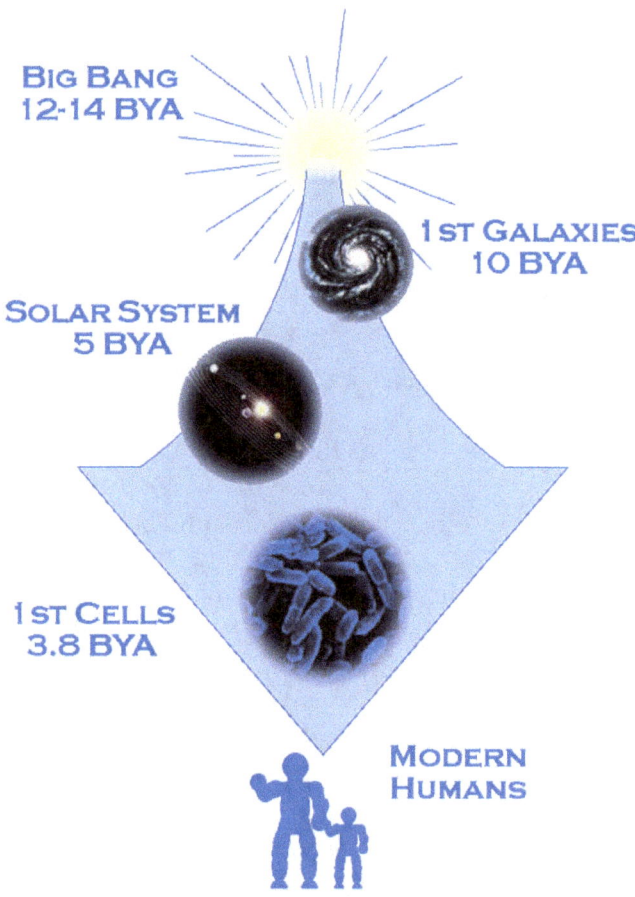

Figure 4.8 Science provides us a universal model that begins with an incredibly small point called a singularity. This singularity is believed to have exploded through a process called the Big Bang into what we now know as the universe. BYA = Billion Years Ago

in a significant way. How can a current event, after all, be connected to an event that occurred many billions of years ago?

The timeless singularity that I describe in this chapter contains all "pasts", "presents", and "futures" simultaneously and this must include the singularity that in our view occurred at the beginning of the universe. The Big Bang was not just billions of years ago but is part of every here and now.

In other words, our present moment is always intimately connected to this universal source, this moment of ultimate creation. Every moment in time is being created by this source now, even though from our perspective this point of creation seems to exist in the distant past.

Moreover, understanding that the singularity at the origin of the universe and the singularity that includes all light are one and the same may explain the effects on mass and energy in Einstein's Special Relativity equation:

$$E = m_0c^2 / (1 - v^2/c^2)^{1/2}$$

This equation predicts that objects acquire infinite energy and mass as they reach light speed. This is normally interpreted as representing the amount of energy that would be required for an object to attain these speeds and is used as an argument against the possibility that any object with mass could ever go that fast. Consider the possibility that this equation is a description of the transformation that occurs as you draw near this state. As you approach this incredible speed (as does all light), you literally join with and become the singularity that contains the entire universe. All the energy that makes up the hundreds of billions of galaxies, solar systems, gas clouds, and even space and time exist within this singularity. You achieve immense energy when you reach this state.

Stop and consider this because it reshapes how we think about the timeless, spaceless state we are discussing. By understanding that the energy of the entire universe is associated with this spaceless and timeless singularity, we can clearly see this singularity is not trivial or something to be disregarded.

We have limited our discussion to the physics of light. However, if electromagnetism and electromagnetic radiation are the physical surrogates of consciousness, then all consciousness should collocate within this same singularity. What we are describing here is a state that contains all consciousness throughout space and time. We can begin to get an intuitive sense for the density of consciousness within this state by considering the 8 billion individuals currently existing on our planet. Now add to that the consciousness of those from the far distant past all the way through to the far distant future. Still, we are considering only a small fraction of the whole since we need to extrapolate to all forms of consciousness that exist throughout our galaxy and from there the entire universe. The density of consciousness in this singularity is immense beyond imagination and provides us a glimpse of Universal Consciousness.

Take a moment to absorb these profound concepts. Special relativity teaches us that two perspectives that seem to be very different can both be valid. Your perspective of existing here and now as a human and the perspective that all "here and nows" are one may both be true. The difference being that the former view leaves you separate and apart from all that is around you, and the latter connects you intimately to all forms of consciousness throughout time and space.

In the next chapter, we will explore further into the nature of Universal Consciousness, contemplating its implications and diving deeper into the mysteries of our existence. Get ready to expand your understanding and challenge the limits of what you thought was possible.

5

THE DANCE OF SYNERGY

Synergy is a central theme of the universe; it runs through everything you can see or touch. This concept gives us great insight into the essence of all things and is required learning for anyone who seeks to understand reality through science.

The concept is that pieces/parts come together to form wholes that are much more than the sum of their parts. Imagine the intricate dance of molecules within a living cell. Each molecule - proteins, nucleic acids, lipids, and carbohydrates - plays its own role, and individually, they are not considered alive. But when these molecules come together in a precise pattern, a miraculous transformation occurs - life emerges!

The whole becomes far greater than the sum of its parts, showcasing

the beauty of synergy (Figure 5.1).

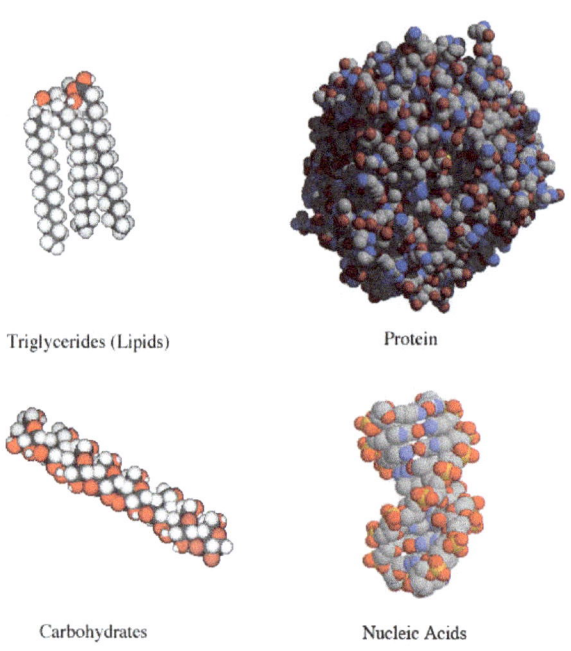

Triglycerides (Lipids)   Protein

Carbohydrates   Nucleic Acids

Figure 5.1 Cells contain four basic types of molecules: lipids, proteins, carbohydrates, and nucleic acids like DNA.

Although synergy provides a powerful framework for us to interpret

the universe, it has been difficult for many to grasp the concept. Our

understanding of living organisms is a great example. Many used to

think there was something different about the stuff that makes up

living systems. It made no sense that you could take something that was as simple and "lifeless" as a bunch of molecules and organize them into something that was living. There had to be more to the story.

These doubts eventually led to the idea of vitalism. Vitalism claimed there was something special about the organic molecules found in living organisms. These organic molecules were somehow imbued with a special life force not found in their inorganic cousins. Thus, a line was drawn between the organic and inorganic world.

These ideas persisted until the German scientist, Friedrich Wöhler, found he could make organic compounds from inorganic reactants (1828). Friedrich Wöhler mixed some inorganic compounds in hopes of producing another inorganic compound ammonium cyanate. To his great surprise, he made a compound called urea, which is the main waste product that we release in our urine.

Wöhler's work clearly demonstrated that organic molecules could be synthesized from inorganic compounds and the distinction between these two classes of compounds became irrevocably blurred. Today we know that inorganic and organic molecules are all made of the same materials – atoms. The difference we see among molecules arises from the number, type, and arrangement of those atoms in the molecule.

Emergent properties arise when new levels of complexity are attained. Life, for example is an emergent property that arises when we put molecules together into the correct complex pattern. It is not something that can be predicted from the component parts, and it arises only at the level of the cell. It seems that every level of complexity has its own emergent properties which operate at and define that level.

When we study the living cell, we acknowledge that this unit has a set of properties that operate at this level of wholeness. We stop thinking of the cell as separate parts. For example, we can study the rate at which cells reproduce or their overall appearance in the microscope (size, shape, etc.). When we study cells at this level, the pieces that make up the cell seem to fade into the background.

The pointillist Georges Seurat created brilliant scenes by placing points of color next to each other on a canvas. When you look at an example of his work, you can see the individual dots of color, each distinct from the next. However, at a distance the points combine to form a picture that is complete and sublime. This is a wonderful example of how individual units can come together to form something on a grander scale that has meaning and significance at that level.

Our human visual field, like a scene within a Seurat painting, is an emergent property that arises from the many qualia produced by our brains. For example, the color we see in our world is produced through the interaction of color qualia. We know this is true because we determine color based on several factors, including the relative shading of all the other colors we see in our visual field. This is called color constancy and it allows us to see objects as the same color independent of the level of lighting. Alternatively, due to these same factors, the exact same color can appear modified under different shading conditions. (Figure 5.2).

Figure 5.2 Color constancy is an important phenomenal experience that allows us to see colors as the same under different lighting conditions. Prove this to yourself by comparing the middle square on the top of this cube to the middle square in the front face of the cube. These two squares are the same color but look quite different due to your expectations based on shading.

Shape information is produced from a set of qualia representing lines of specific orientations. Bits of lines are brought together to form line segments and then shapes. At a higher level, color and line/border information is brought together to form the objects we see in our world like the cube in the figure above. It is easy to see how this continues until we reach the level of our entire subjective experience.

Synergy is a concept that goes way beyond biology. It is a quality that runs deep into the essence of all that we consider to exist. As far down as we can "see", parts are coming together into networks to form larger substances with new and unpredictable properties. Quarks come together to form subatomic particles like protons. Protons, neutrons, and electrons come together to form atoms. Atoms form molecules that come together to form cells, and cells form tissues and organs. This process continues until it reaches the level of organisms like us.

Naturally, this process does not stop with organisms like humans but continues on up. Scientists have identified higher levels of organization like populations, communities, and ecosystems, each with their own rules that apply at that level. For instance, the evolution of organisms occurs at the level of populations and not individual organisms. You know this is true because you yourself are not physically evolving. However, the population of humanity is evolving right now.

Scientists have spent a lot of time defining the properties of these higher levels. Interestingly, we are told that the properties of life and consciousness end with us. None of these higher levels are considered to be a living entity, nor are any of these higher levels believed to display any form of consciousness. Why is it that we have come to the conclusion that we are the highest rung in the ladder when it comes to life or consciousness?

Let me use an example to make the point. I am going to pass you through my incredible shrinking machine and take you on a short journey. There, you are now the size of cell. Come over here and talk to a few cell friends of mine. If they could think, wouldn't cells come to the same conclusion that we have? From a cell's perspective, the highest-level worth considering would be that of the individual cell. After all, organisms like humans are only a home to these cells.

Our bodies are places that are maintained so cells can continue to live and reproduce as they should. When a new home needs to be made, it is individual cells that come together and go through the trouble of reproducing themselves many times over to create a new body they can again call home. If there were such creatures as scientist cells, they might never even consider that they themselves were contributing to some larger form of life or consciousness.

It is natural to slant things in a way that makes whoever is telling the story the center of all that is interesting. Let us try something different and take off that bias. What if instead of being at the top of the ladder, we were to find ourselves somewhere in the middle? What if, like the cells inside your body, we are also contributing to a larger existence? These ideas are not new. Many cultures of the past considered the earth to be alive. The ancient Greeks considered the earth to be a living entity and referred to it as the mother goddess Gaia. The biologist James Lovelock revived the idea in modern times with the Gaia hypothesis—the biosphere as a living organism.

Lovelock believed our planet has an intricate physiology that spans the entire globe. Components of this physiology include grasses and microbes, animals and humans. Inorganic materials like seawater are also thought to contribute to the structure and function of this living form. This would make our biosphere the largest organism in our solar system. Although the idea may sound foreign, it is actually not so strange.

As a biologist, it is a small extrapolation for me to see a much grander metabolism existing throughout our planet. The metabolism that occurs within your body includes a coordinated set of chemical reactions called metabolic pathways. Why should we artificially separate the metabolic pathways that can occur within an organism from those that occur through the action of several organisms?

Nitrogen provides an elegant example of this concept. Nitrogen exists in large quantities as a gas in our atmosphere. This is fortunate for us because nitrogen is also in high demand by our bodies because it is used to form biological building blocks like amino acids and nucleotides. However, despite the huge supply and the great demand, there exists a major problem. We cannot use nitrogen in its gaseous form.

Soil microbes to the rescue! Many of these microorganisms develop symbiotic relationships with plants and convert nitrogen gas into useful forms like nitrate. Legumes (beans and peas) are examples of plants that have these types of symbiotic relationships. The plants take the converted nitrogen from the microbes and use it to make amino acids and all the other types of nitrogen-containing biomolecules required for life. Animals ingest plants to acquire the necessary nitrogen, and humans eat other animals and plants to get these nitrogen-containing molecules. As you can see, there is a defined set of chemical reactions necessary for processing nitrogen. These reactions do not belong to any one specific organism but are carried out by several, each making its own contribution. The chemical conversions we see in nitrogen processing are of the same type as the multi-step chemical conversions that occur within your own body. The nitrogen cycle is just one example of many "planetary metabolic pathways".

Our planet has other qualities of living organisms. An important property of life is that living systems can regulate their internal environments. Interestingly, our planet regulates its temperature within a certain range, maintains the concentrations of atmospheric gases within a certain range, and maintains the salinity of its oceans within a certain range. This is extremely important for humans because variations from these ranges would surely be disastrous for us.

Another quality of living organisms is that they reproduce. As I write these words, we have created an artificial living environment in space—the International Space Station (ISS). This space station allows fellow humans, plants, and animals to live in space, apart from earth. Albeit, it is not as comfortable as mother earth, but still the ISS allows for life. Could we consider this a rudimentary type of planetary newborn?

If that concept doesn't work for you, many people are talking about using the space station as a stepping-stone to Mars. There has been much interest in recent years in the idea of terraforming Mars—trying to make Mars habitable by humans and other organisms. Maybe, maybe not, but if by the next millennium Mars becomes habitable through our efforts with a blossoming population of earth creatures couldn't we say our biosphere has reproduced itself?

Why did I bring this whole synergy thing up in the first place? Was it to argue for the Gaia hypothesis? I believe the Gaia hypothesis will one day require that we reexamine how we define life, but that is not the reason why I brought up synergy.

Within your brain, you have billions of neural cells that miraculously come together to form your conscious experience. These comparatively simple structures, through their multitude of interactions, allow me to write this book for you. The question I pose to you now is this: Are we the highest form of consciousness? Just as we could be participating in the physiology of a larger organism (Gaia) could we not also be participating in a greater consciousness?

If relatively simple forms like single neural cells can create a consciousness like ours, is it hard to imagine that higher forms of life like humans could contribute to an even greater form of consciousness? Does it seem reasonable to you that in this massive universe of several hundred billion galaxies that we are the highest rung in the ladder when it comes to consciousness?

What would this higher consciousness be like? Most likely it would be aware at a completely different level than you or I. As in the rest of the universe, you would expect synergy and emergent properties arising at this level of wholeness. This consciousness would be qualitatively different from that of a human, simply because it would not have a body as we know it. It would likely have an entirely different concept of time and space.

Many religious traditions of the past point to this consciousness. To grasp something so large, we in the West have used the term God. Taking the lead from these well-established traditions, we can glean more. This Universal Consciousness informs our lives just as we inform Universal Consciousness. A poor analogy would be the process by which our conscious thoughts activate the neurons within our brains and neurons, through their activity, contribute to our thoughts. The quality of the relationship we have with Universal Consciousness, however, is quite different than the one we have with the parts that make up our brains.

Our entire experience (everything we know, feel, or do) exists within Universal Consciousness. This Consciousness knows us completely, so our lives inform Universal Consciousness. Universal Consciousness also informs us. The information we receive from Universal Consciousness often operates at the level of metaphor. Many of these metaphors are the symbols and stories we find in our significant dreams and societal myths. From this, we see that besides being intimately connected to Universal Consciousness we also can have a real relationship with this higher form of consciousness. And yes, there is much more...

6

A TAPESTRY OF UNIVERSAL CONSCIOUSNESS

In our quest to unravel the mysteries of the universe, we now stand at the precipice of a profound revelation—the Singularity of Light. This is the point where all that we have discussed thus far converges, and it culminates in the awe-inspiring concept of Universal Consciousness. Prepare to venture into the realm of pure experience, where the physical and the mental coalesce to form the fabric of existence.

To comprehend the magnitude of Universal Consciousness, imagine an immense and resplendent tapestry laid before you. This tapestry is unlike anything you've ever encountered, as it is composed entirely of brilliantly colored dots. As you gaze upon its mesmerizing beauty, you can't help but recall the works of the painter Seurat.

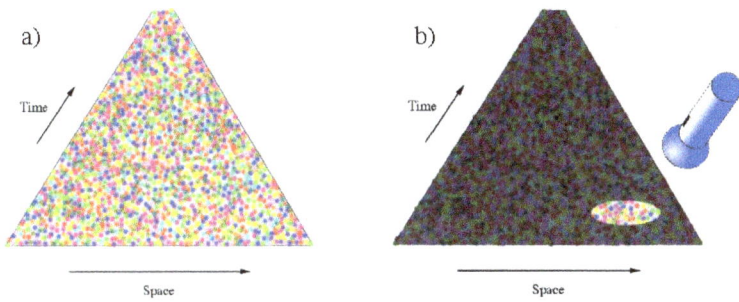

Figure 6.1 a) Universal Consciousness includes all awareness produced throughout time and space. Conscious awareness is indicated by the dots which correspond to qualia produced by at least nervous systems such as ours. b) From this higher perspective, our form of consciousness is produced by focusing on a subset of qualia that together produce an instant of human experience. Our human experience of time is created by moving the circle of light forward through the dimension of time. Within Univerasal Consciousness, however, experience is not separated but remains part of the whole.

In this metaphorical tapestry (Figure 6.1a), we find a two-dimensional representation of the qualia produced by living organisms with nervous systems, such as ourselves. One dimension represents three-dimensional space, and the other dimension is time, encompassing all awareness throughout space-time—from the beginning to the end. The immensity of this tapestry is a testament to its comprehensive nature, encapsulating every conscious experience that has ever been or will ever be.

Our brains produce qualia. These small EM fields are the physical manifestation of consciousness in our brains. The combination of these qualia generates our human consciousness, and the specific set of qualia within our minds shapes our individual experiences. If we were to assemble a different set of qualia, our experience would radically shift.

Now, imagine yourself in this tapestry room, with a dim light hanging overhead. As you hold a flashlight above the floor, you create a small circle of light on the tapestry. This circle contains a particular set of qualia that gives rise to your current conscious state. But what happens when you move the flashlight through the dimension of time?

As you move the flashlight, some qualia that were originally in your circle of light fall out, while new qualia enter the circle. This dynamic composition of qualia results in a continuously changing experience of reality. Our human experience of time is created by this moving circle of light (Figure 6.1b), allowing us to perceive reality as a sequence of moments.

From the perspective of Universal Consciousness, all qualia coexist simultaneously, transcending the limitations of time. In this grander state of awareness, the entire tapestry is visible all at once, and experiences are not separated by time but remain part of a timeless whole. There is no past or future; all sorrows, joys, and experiences converge into an eternal instant.

The ecstatic experiences of the many that have achieved this state of wholeness through prayer, fasting or meditation, are just this. These individuals were able to dim the light of their flashlights at least momentarily. By doing this, they perceived much more of the tapestry and were astounded, overwhelmed by the beauty of what they felt and saw. Universal Consciousness is ever present; our "eyes" just need to be opened to it.

Within Universal Consciousness, you are never alone. You are part of an integrated whole, connected to all conscious beings across the universe. Other lives coexist with our own, existing as part of this grand tapestry of Universal Consciousness. Universal Consciousness is a boundless source of knowledge. By existing in this state, you gain conscious access to all the information contained within it. This access is not limited by space or time, as the singularity represents a timeless "here and now." Phenomena like psychic abilities and sensing future events become comprehensible when seen through the lens of Universal Consciousness.

Love takes on a transcendent quality within Universal Consciousness. I previously shared that love is the way we sense deep heartfelt connection. Universal Consciousness is the ultimate point of connection for all consciousness throughout the universe. It is a state of being that exceeds any love experienced on Earth. This state of bliss and connection is accessible to all of humanity. Those who have had near-death experiences often describe being enveloped by a love beyond anything they have ever known in their earthly existence.

Universal Consciousness is not an abstract or dissociated energy field—it possesses consciousness and intentionality. Just as we have a sense of self at our consciousness level, Universal Consciousness has agency and fully understands the relationship it has with us. It communicates with us through dreams, visions, and experiences, while we can communicate with it through prayer.

What I have written here overlaps with the beliefs of many religions. I am purposely trying not to use terms from established religions because I am developing a scientific understanding of this transcendent state of consciousness. This scientific understanding is not a replacement for religious beliefs but is meant to add to and deepen the understanding of our spiritual nature.

I believe what I describe here is real and an important part of ourselves and the universe in a scientific sense. It has the potential to illuminate all other aspects of scientific understanding by revolutionizing the way we understand reality. Whole new undiscovered landscapes of science will open for those that wish to explore this part of the universe. The new perspectives that arise from this exploration will allow science and our societies to grow in ways that are important for our mental and physical health. It will allow us to gain a greater understanding of our connection to nature and the role we play in maintaining the health of our planet. Most importantly, it will allow us to clearly see that we are in each other's heart.

In the pages of this book, I've ventured through myriad perspectives in an attempt to unveil the transcendent nature of Universal Consciousness. Each exploration has nudged us closer to understanding, yet its elusive essence continues to slip just beyond our grasp.

Quantized Visual Awareness is a model that unveils the mechanics behind the creation of qualia within the human brain. Here, qualia exist within a timeless dance, entwined with the very fabric of light itself. A useful metaphor is the breathtaking tapestry of awareness, woven intricately across space and time, each thread a representation of consciousness in its purest form. At the heart of this cosmic tapestry lies the singularity of light—an awe-inspiring point of connection where light and consciousness converge into a singular source. It is from this sacred nexus that the universe was born, its creative energy pulsating through every moment of existence.

As we journey further into the chapters that follow, we will delve into the empirical evidence supporting the existence of Universal Consciousness. We will explore psychic, near-death, and past-life experiences that have brought individuals closer to this transcendent state of conscious connection. By the end of this exploration, you may decide the evidence for Universal Consciousness to be overwhelmingly compelling.

Now, join me as we embark on a path of enlightenment, seeking to comprehend the very essence of existence—the Singularity of Light and the boundless tapestry of Universal Consciousness that awaits our exploration.

7

EXPLORING THE ENIGMATIC REALM OF PSI

Let us enter the fascinating world of PSI—a domain where our consciousness converges with the profound tapestry of Universal Consciousness. In the previous chapter, we explored how our thoughts and experiences are integral threads woven into this vast cosmic fabric. Could it be possible that through our common connection to Universal Consciousness, we might also be connected to each other?

Our normal perception often confines us to a limited awareness, tethered to the present moment. However, as parts of a continuous whole, we possess the innate ability to extend our consciousness beyond our typical boundaries. Imagine expanding your awareness into distant realms of the tapestry—regions beyond your usual identification. This extraordinary expansion of consciousness is akin to developing new perceptual abilities—a heightened form of sight, hearing, and communication, beyond the ordinary sensory experiences.

These extraordinary abilities collectively fall under the term "PSI" (Greek letter ψ). From clairvoyance and clairaudience to telepathy and precognition, PSI embraces a wide array of extra-sensory perceptions that defy our conventional understanding of reality.

You might be wondering if these phenomena are merely anecdotal or have been subjected to rigorous scientific scrutiny. The truth is, there is ample evidence supporting the existence of PSI, often experienced between individuals with close connections. For instance, someone feeling an urgent need to check on a loved one, only to discover that their intuition was correct, showcases a form of clairvoyance.

PSI events occur much more frequently than we realize in our society. Whenever I have taught a course that covered the topic and asked if anyone had an experience that would fall under this category, at least thirty percent of those attending would raise their hands. Many have sadly remarked to me that they felt there were few places they could comfortably talk about these sorts of experiences.

In our culture, discussing such experiences is often met with skepticism and hesitation due to the dissonance they create with modern Western scientific perspectives. This reluctance to explore our natural connection to Universal Consciousness can lead to feelings of disconnection and even depression. But as history has shown, assuming you have a complete understanding of reality based on the current level of knowledge is a mistake.

Many scientists and others of the late nineteenth century were convinced that all the important ideas about physics had been discovered and all that remained was to tie up a few loose ends. The irony of this became evident when scientists started playing around with those few loose ends and the entire structure of the field of physics started to fray. This is exactly where Einstein and others came in and gave us a totally different understanding of reality based on quantum physics and strange ideas like special relativity that we discussed earlier in this book.

In many ways, we have arrived at a similar place at the beginning of the twenty-first century. The loose ends we have now, however, are somewhat larger. These loose ends relate to the experiences many have had and for which Western natural science has offered no explanations. I have no doubt that throughout this century more and more will come to acknowledge the reality of these experiences and turn to study these very natural experiences in earnest.

A common misconception is that PSI phenomena are manifesting through time and space. However, the electromagnetic fields produced by our brains are relatively weak and it would be hard to imagine how the subjective experience they carry could be conveyed from one individual to another at a distance. In addition, some of the experiments analyzing these phenomena were carried out in electromagnetically shielded rooms (for example, work done at the Stanford Research Institute), which precludes the transference of electromagnetic radiation.

An important break with much previous thinking about PSI phenomena is this, the electromagnetic fields (and the information they carry) are not traveling through space and time in order to convey information from one individual to another. This information exists simultaneously in our brains and the singularity of light discussed previously. It is our direct connection to this singularity and Universal Consciousness that allows us to access information arising in other brains. The key issue here is our direct connection to Universal Consciousness and not a direct connection to each other.

Through our intimate connection to Universal Consciousness, we become aware of information that lies beyond our normal everyday experience. Some examples include accessing information about our "future" selves through what is called precognition or the closely related déjà vu. The tapestry model I described in the last chapter illustrates the point that gaining information about your "future" self does not require you to leave your body or to travel through "time." Learning about your "future" self means becoming aware of other parts of the tapestry. In other words, we can expand our awareness to include normally unavailable qualia, and this is PSI.

We are woven into the tapestry and cannot be separated from it. This tapestry of consciousness includes all awareness throughout space and time. It is only when we look at it from our traditional human perspective, that it appears as if we are becoming aware of something that lies in our future.

What happens if you expand beyond the part of the tapestry, we normally associate with ourselves? If you access a part of the tapestry you think of as someone else, we consider this to be telepathy. Telepathy is defined as the ability to communicate with others through thought and without using the five common senses. However, we are not really communicating with "others"; again, you are just becoming aware of another part of the tapestry with which you normally do not identify. To us, due to our limited understanding of Universal Consciousness, it appears that we are communicating across some distance.

Many are surprised to find out that PSI has been studied rigorously. Researchers have gone past anecdotal experiences and scrutinized this phenomenon using accepted scientific protocols at major research institutions. Through this research it, has been well established that PSI does exist and is innately part of our human natures (read *The Conscious Universe* by Dean Radin for a more detailed description of these studies).

In setting up these experiments, researchers look for significant differences from the null hypothesis. For example, if I had five different types of cards and I asked you to guess which card I am holding, you should be right 20% of the time due to chance. Based on this frequency, the null hypothesis for this experiment would be that you should guess correctly with a frequency of 20%. When these types of experiments are carried out, scientists often find that those guessing will do better than the predicted 20% on average.

In science we like to test the reliability of the results we see. It is possible that someone may guess correctly more often than expected just due to random effects after all. To address this issue, scientific research uses statistical methods to determine how confident the researcher can be about the results and ensure the observations do not arise due to random fluctuations. This is done by determining the 95%

confidence interval. This is the range of values around your data point that allow you to say with 95% confidence includes the true value of the data point. This confidence interval is heavily dependent on the number of times you repeat the experiment.

Figure 7.1 A comparison of two data sets, generated either randomly by a computer or by querying a participant of the study. Given that the 95% confidence intervals (shown by vertical lines) do not overlap, we can say these two data sets display a significant difference.

In the example above, we could say with confidence that we have disproved the null hypothesis (the results are due to chance) if we see an average correct guessing rate of 35% with a 95% confidence interval of +/- 5% (Figure 7.1). This range is far above the predicted 20% with a similar 95% confidence interval and indicates something other than random guessing is occurring here.

Of course, this is not the end of the story because we need to ensure the experimental design includes checks against the transfer of information between the experimenter and the subject that is guessing. Blind techniques are standard across science and are used in research as diverse as pharmaceutical drug trials and PSI research. Subjects are separated from the researcher and in rooms that are shielded (acoustically, visually, electromagnetically) from each other. These experimental designs remove any doubt that the experiment was contaminated either consciously or unconsciously by the researcher.

Given all these safeguards, much PSI research shows correct guessing averages that surpass the expected rate due to chance. However, it is also observed that in some experiments the 95% confidence interval overlaps with the expected value due to chance. It is likely these contradictory results arise from the varying PSI ability of subjects participating in these studies. Like any other skill, you would expect some individuals to excel in PSI while others have a limited capacity.

Most recently, due to the limitations of many research projects (time, resources, people) a new more comprehensive approach to analyzing the results of scientific research has become common and this is called meta-analysis. This is an analytical research tool used to integrate the data from many research projects into one study. Through these techniques, developed over the past several decades, it is now possible to truly see if the results seen in various experiments point to a deeper truth that encompasses them all.

Meta-analysis has been used to study the efficacy of pharmaceutical drugs when the outcomes of individual experiments have not been clear. It is also being used to look at a good number of PSI studies. Meta-analysis has further strengthened the case for PSI. The consistently observed differences between experimental observations and chance predictions cannot be dismissed, indicating a genuine phenomenon.

You may be wondering at this point about the specifics of the experiments that have been carried out over the past 100+ years. Yes, the guessing of card types has been a significant part of these studies, but these were more common in the first half of the twentieth century.

Since then, the experiments have become more specialized, and even more security focused to address the concerns of skeptics.

Experiments on telepathy are now designed to decrease the amount of peripheral stimulation subjects receive during experimental trials. In the 1960's and 1970's, research was often done with sleeping subjects that were asked to report on their dreams to see if they were prone to pick up on PSI information during their dream state.

From the 1970's on, researchers began using the Ganzfeld method to investigate PSI phenomena. These experiments focus on limiting the receiver subject's sensory experience by using white noise and eye covers. The receiver is asked to talk about whatever images they are perceiving while a sender contemplates an image, object, or video that is the target of the experiment. Besides sensory stimulation, these experiments are much more freeform than the earlier card experiments since receivers are allowed to discuss whatever images come into their minds.

Through rigorous investigations like these, it appears that PSI is a real phenomenon. The blind-experiments and use of statistical analysis (including meta-analysis) are the same type used throughout science. If we approach this topic without bias, then we must conclude that that the consistently significant differences between experimental observations and the chance predictions based on null hypotheses cannot be disregarded or tossed aside.

While skepticism and societal pressures may have obscured the recognition of PSI's reality, it aligns perfectly with our direct connection to Universal Consciousness. PSI is not about the transference of information through space and time; instead, it is our access to information that exists within Universal Consciousness in the "Here and Now". By expanding our awareness, we tap into normally unavailable qualia, allowing us to glimpse the tapestry beyond our usual limitations.

Through quieting the mind and reducing sensory stimulation, we can facilitate the access to PSI information. Meditation, prayer, and sensory deprivation experiments like Ganzfeld can aid in connecting with Universal Consciousness and its vast array of experiences.

PSI provides compelling evidence of our intimate and continuous connection to Universal Consciousness. Embracing this notion revolutionizes our understanding of ourselves and the universe. While PSI is a significant support for the existence of Universal Consciousness, there is more to explore, which we will unravel in the next chapter.

Now, join me as we continue our journey of enlightenment, peering deeper into the mysteries of Universal Consciousness and its profound implications for our lives. Open your mind to new possibilities, for what awaits us is nothing short of awe-inspiring!

8

REVELATIONS FROM BEYOND: NEAR-DEATH EXPERIENCES

Near-Death Experiences (NDEs) are quickly becoming a field of study that has the potential to transform our understanding of ourselves and the universe. This fascinating domain emerged with recent advancements in medical technology, allowing the resuscitation of individuals who experienced clinical death due to severe trauma or illnesses. Thousands of these NDErs have shared astonishing accounts of their experiences while being clinically dead, which defy our conventional notions of reality.

These experiences are far from vague dreams; they are described as super-vibrant and incredibly vivid, often more real than everyday waking life. Some skeptics argue that NDEs are mere hallucinations of a dying brain, but here's the catch—many NDEs occur when the brain has little to no neural activity and is deprived of oxygen for extended periods. How could such lucid experiences be produced by a brain in such a state?

Dr. Jeffrey Long, in his book *Evidence for the Afterlife*, presents compelling scientific evidence supporting the reality of NDEs. Patients in these conditions often experience hypoxia—a state with reduced oxygen levels that should lead to confusion, memory loss, and fatigue. Yet, NDErs recount experiences that are clear and coherent, which contrasts starkly with this hypoxic state.

To understand this phenomenon, consider the analogy of a flashlight illuminating the tapestry of qualia (from Chapter 6). When we die, it's as if the flashlight is turned off, and the parts of our brain responsible for immediate, vivid experiences cease to function. We transcend the usual limitations of our personal experience, expanding to a larger sense-of-self with access to a broader spectrum of experiences beyond our usual perceptions.

Death does not erase experiences; they persist in the timeless state of the tapestry within the singularity of light. Our entire life—past, present, and future—exists in an eternal instant within the singularity. After death, we continue to exist solely within this eternal instant.

One of the common elements in NDEs is the out-of-body experience, where individuals find themselves detached from their bodies and can observe the surrounding environment. Remarkably, they report details that were inaccessible to them physically, like witnessing objects in unusual places or overhearing conversations in distant rooms. These reports serve as evidence of the authenticity of NDEs.

A well-documented case is that of a patient Maria (please see *The Near Death Experience: Problems, Prospects, and Perspectives*; edited by Bruce Greyson and Charles Flynn) who suffered a severe heart attack in Seattle Washington. In her out of body experience, she noticed a shoe sitting on the ledge of the third floor of the hospital. The shoe was described as a man's left-tennis shoe, colored blue and with a shoelace tucked under the heel. Since this was the third floor, it would have been impossible for the patient to have seen this from ground level. The truth of these statements was investigated by the researcher Kimberly Sharp, who took it upon herself to look out of the third-floor windows of the hospital. She found the shoe exactly as it had been described.

A more recent case is given to us by Dr. Mary Neal an orthopedic spine surgeon. She was kayaking in South America and was stuck under water as her kayak was thrust down below the frothy surface by the heavy flow of water off a waterfall. She drowned, and then saw the frantic scene created by her companions as they searched for her over the river.

Her descriptions of how she was found and unconsciously taken from the scene to an ambulance (a miracle in and of itself) are described in her book *To Heaven and Back* and coincide precisely with the experiences of her companions. Given that her body was underwater as the rescue began, it would have been impossible for her to know the specifics that occurred as her friends and husband first realized she was missing and were intensely searching for her at the surface of the river.

There are literally thousands of these types of reports that are well documented and studied by researchers in this field. This massive amount of data is significant and is difficult to dismiss. However, the question arises as to the mechanism by which these experiences occur. It is hard for a scientist to accept observable phenomena that contradict or surpass the current level of scientific understanding of the universe.

It is possible that these perspectives arise out of the inherent consciousness embedded in the electromagnetic fields that are all around us. Perhaps, as we first transition out of life, we access this conscious field that normally surrounds us. In doing so, we gain information that cannot be coming directly from the deceased person's physical sensory systems. In the end, these out of body experiences may provide support for the existence of this very natural conscious field.

Those that have recounted their near-death experiences often claim that after their initial experience of seeing their lifeless body, they begin to move through a tunnel or other space at incredible speeds. This seems like a necessary change in moving from our mass-based reality to the state of the singularity of light. As their consciousness shifts from our material, earthly state to this new state, it must be brought up to light speed so that they can join the singularity with all other light.

As they reach this state, time stops as is the natural condition for all light. Accounts differ in the amount of time that people claim they spent in their NDE, but there are accounts that seem to last years even though the time here, in our world, is much shorter. Time doesn't exist in the NDE state and does not correlate with the passage of time in our material state.

Life reviews are another intriguing aspect of NDEs. In many cases, NDErs claim they re-experience moments from their lives, but with a twist—the emotions not only of themselves but also of others affected by these events. This allows them to gain a much deeper understanding of how their actions ripple out to others.

NDErs describe these life reviews as something like watching a movie of their life. This is consistent for individuals across cultures. Given the cultural variation that exists in understandings of the afterlife, it is more than coincidental that these life reviews occur in the movie-like fashion. Also, given that many skeptics argue these experiences are dream-like, hallucinatory phenomena, it is more than odd that the life reviews are lucid, 3rd person movie-like experiences with heightened knowledge of the emotions of all the participants of the scenes that are seen.

How do these life reviews fit within the model we are developing? As stated previously, the tapestry contains all experience. We could describe the same process by saying that the NDEr is allowed to experience the parts of the tapestry that correspond to their life but also to parts that correlate with the life of others. In other words, the NDEr is being allowed to experience a greater swath of this tapestry and thereby see how their life directly and indirectly affected the life of others.

Those that report near death experiences often find themselves enveloped by light. NDErs describe environments filled with light and objects that appear as if they are made of light. People and objects themselves shine out their own light or act as conduits of light. The fabric of the clothes people wear look as if they are weaved from light. In some cases, there are reports of encounters with individuals that emit so much light that the NDEr is surprised they are not hurt by this much light. They seem to realize that in life this much light would damage their retinas.

There is no coincidence here. Afterall, we are talking about the singularity of light. It makes sense that the very fabric of the materials and beings is light itself. I believe these first-person descriptions greatly support the idea that all light comes together into a singularity of light and that this state is timeless in nature.

Moreover, this light is not like any we have ever experienced. It seems to be laden with love and peace. I have shared with you that love is how we experience connection. Given that the singularity of light is the ultimate point of connection it also makes sense that the light itself should correlate with the experience of love.

The research and personal accounts in the field of NDEs are vast, with numerous rigorous studies and reports in scientific journals. I encourage you to explore this fascinating topic yourself, and you'll likely be amazed by the wealth of evidence supporting the reality of these profound experiences.

As you venture deeper into the enigmatic territory of NDEs, be prepared to question and expand your understanding of life, death, and the boundless mysteries that lie beyond our physical realm. The revelations from these near-death encounters will challenge your perceptions and unveil the profound nature of our existence.

9

REINCARNATION

The world of reincarnation is a tapestry woven with threads of consciousness, interconnecting our souls in ways we can barely fathom. While Western beliefs tend to emphasize individuality, Eastern philosophies, like Buddhism and Hinduism, embrace the idea that we are all part of the Absolute, experiencing a form of energy that is reborn into new lives. This fascinating concept challenges our conventional views of separateness and independence.

Reincarnation has become a hot topic in the media. Despite not being native to our culture, it has captured the imagination of many. As a result, researching this phenomenon objectively in the West has become challenging, as skeptics argue that stories may be tainted by exposure to widespread media coverage.

However, reincarnation studies in the West began long before reincarnation became popular in our culture. One example is the work of Dr. Ian Stevenson who worked at the University of Virginia School of Medicine for several decades and founded the Division of Perceptual Studies at the University of Virginia. Dr. Stevenson focused his work on children, since he believed they would be less contaminated by social norms and popular culture. Indeed, over his entire career, he researched approximately three thousand cases of claimed reincarnation amongst the young.

In these thousands of cases, Stevenson uncovered several characteristics that typified these cases. I will list five of these here. Most cases don't include all five but only some of these characteristics. The five are simply the most common attributes associated with these cases.

Firstly, comparisons of many cases often found there was a prediction of reincarnation by the individual who would eventually die and be reincarnated. These predictions included details of the family into which the individual would be reborn and specific markings to be found on the body of the child.

Another common characteristic of these cases is the announcing dream to the mother or relatives of the child. This can be as explicit as the subject indicating their wish to be reincarnated in dreams of the future mother. These dreams occur, with few exceptions, before the birth of the child.

A third common property of these cases is that there will often be birth marks or birth defects on the child at the same body location where wounds or traumas occurred to the individual that died. Stevenson corroborated this, when possible, by looking at autopsy reports of the subject that died. Recorded stories from friends or relatives of the deceased subject with notations about traumas and injuries of the deceased were also used as evidence in these cases.

A key component of these cases is the claims made by children about previous lives. These claims contain details that could not be known by the young child and often not even by his or her parents. These details include information about events, individuals, and objects in the deceased person's life.

An example of this is given by the popular story of the current Dalai Lama, Tenzin Gyatso. In the Tibetan Buddhist tradition, it is believed that the Dalai Lama is reborn as the next Dalai Lama and Tenzin Gyatso is thought to be the 14th incarnation in this long line.

When he was a boy, Tenzin Gyatso was approached by Buddhist monks and shown several objects, some of which belonged to the previous Dalai Lama. It was noted that he only picked the objects that belonged to the previous Dalai Lama and left the other items untouched. In his memoirs, the Dalai Lama remembers that he recognized one of the monks in the search party even though he was dressed as a servant and then asked for the prayer beads worn by the monk. These beads also belonged to the previous Dalai Lama.

Beyond the specific claims made by children, a common characteristic in these cases is the odd mannerisms and speaking styles of these children. Children developed mannerisms and ways of behaving that were not being modeled for them in their homes yet seemed very similar to mannerisms recorded for those who had departed.

According to Dr. Stevenson, out of 245 cases in India, the average age at which children began speaking about previous life experiences was 38 months (all in the range of 2-5 years old). This was the same for children from other cultures including the United States.

The fact that these common features are seen amongst people of widely diverging cultures (including cultures that don't believe in reincarnation), argues for the reality of these phenomena independent of cultural factors.

There were certain requirements that were in place to ensure the cleanest and highest quality examples in these studies. For example, when a child claimed to know information about a previous life, but no pre-existing individual could be identified, the information was discarded. All studies included interviews with multiple firsthand informants, which were repeated to ensure consistency from telling to telling. Independent verification was required of those who knew the previous individual to determine if the information the child was describing was accurate. All pertinent documents (medical records, postmortem examination records, etc.) were located and copied.

Work in the field, led Stevenson to develop a questionnaire which served as a reminder of what information was to be determined in every case. The types of information the researchers were trying to glean fell into several categories. These included: what the child communicated about his or her previous life; what possibilities existed that this information was inadvertently communicated to the child (for example, did the families know of each other or had the parents heard reports of the death); noting any unusual behavior that seemed to echo a behavior of the previous life; details of the mother's pregnancy and the early development of the child; other pertinent information like the child's birth order and evaluations of his/her intelligence.

When possible, interviews were recorded by two individuals and audio recording was used when interviewing the child. Although interpreters were often used, Stevenson used interpreters that were reliable, and he reused the same interpreters since they would learn the terminology and understand the questions more clearly with experience.

Dr. Stephenson adopted well-established methodologies used in other fields of study. He adhered to high standards of requiring accuracy and details in his cases and required checking and cross-checking when possible. He only pursued cases that he deemed reliable such as only focusing on young children who were less likely biased by popular culture. In addition, he only accepted cases for which a previous life could be explicitly established.

Furthermore, he pursued cases in cultures, like those of North America, wherein reincarnation was not well accepted and less likely influenced by societal factors. Taken as a whole, given the large number of well-documented cases in various cultures, these studies indicate reincarnation does occur.

I am including a specific case taken from Dr. Stephenson studies. This is being shared with you for you to consider how compelling and detailed a specific case can be. If you would like to learn more about this case as well as many others, please read his well-known book *Children who Remember Previous Lives* by Dr. Stevenson and published by the University Press of Virginia.

The caste system of India is well-known. There are various well-defined levels that correspond to one's position in society. The top of this caste system is that of the Brahmins while the lowest caste is known as the untouchables. Jabir Singh was born into a low caste family in India. However, this boy remembered a high caste life of a Brahmin. He looked down on the members of his family since he perceived them to be of a lower caste and he shared this with them. This sense was so strong that he often refused to eat their polluted food and found himself starving. Naturally, this created conflict with members of his family and the behavior was not rewarded but often punished. As he grew, he had trouble finding work since he felt the menial nature of the work he was offered was beneath him.

It is easy to grasp that this child did not receive encouragement to act in the peculiar way he did. Being a low caste family, his relatives had little contact with those of the Brahmin caste so there was little opportunity for these behaviors to be learned. He was also punished for his behavior so there was no reinforcement for the behavior even if it was learned. His state persisted for years and thus we can suppose that he identified strongly with his Brahmin past in the face of much negative feedback. This case clearly shows that claims related to previous lives arise in often unreceptive conditions with little incentive to continue this behavior unless it is truly heartfelt.

Dr. Stevenson did not conclude that reincarnation was a universal process. Indeed, he made the point that a significant number of these cases arose when a violent and sudden death had occurred. Also, it can be argued that some of these observations arise from PSI abilities as we discussed earlier and do not indicate a reincarnation process per se.

Reincarnation challenges the Western emphasis on individuality, revealing the intimate connections we share. It suggests that our lives may be part of one continuous thread, woven intricately through what we consider separate lives. Such understanding only makes sense when we acknowledge a realm beyond the traditional confines of our material life, a realm where Universal Consciousness connects us all in ways beyond our comprehension.

Through reincarnation studies, near-death experiences (NDEs), and PSI studies, we glimpse the grand tapestry of conscious existence, transcending the boundaries of our individual lives. These studies imply that we are entwined within a vast consciousness, deeply connected to each other beyond our closest relationships. It's an intimate bond that blurs the lines between separate lives, revealing the profound essence of all consciousness—that we are all connected and part of one Universal Consciousness.

10

I AM

We began our discussion by looking at consciousness and how it might be understood in terms of science. We know that the human brain can shape or produce subjective experiences like vision. Evidence comes from the many studies of individuals that have suffered lesions to their brains and lost the ability to see, hear, smell, or taste.

Gazing into the complex world of the brain, we encounter three potential candidates for the physical aspects of consciousness: the neural cells themselves and the molecules that compose them, the electrical currents flowing through the neural circuits made from these cells, and the electromagnetic (EM) fields produced by these electrical currents. Among them, I believe the EM fields hold the key to consciousness, as they automatically integrate local information into the brain's larger EM field. This enables any part of the brain to contribute to the overall conscious experience instantly as it becomes active. Moreover, our constantly changing and dynamic conscious experience correlates well with the rapidly changing EMF of our brains.

One of the guiding principles from biology suggests that biological systems don't invent new physical properties; they utilize existing ones for survival and success. This concept extends to consciousness and suggests that a fundamental and subtle form of consciousness, potentially rooted in electromagnetic fields, existed in nature long before it evolved into the diverse expressions of conscious experience seen in the animal kingdom today.

In humans, we find that our conscious experiences are dense with sensations, emotions, and even thoughts. This has led us to the mistaken impression that consciousness is the property of individuals instead of a property of the universe. We have fallen into the trap of thinking that consciousness only arises in our brains and that this conscious experience belongs only to us and not to the universe. This form of thinking also leads to the mistaken belief that no one else can access our inner subjective experiences and so psychic experiences that go beyond our personal experience cannot be real. In addition, the belief that our conscious experience is a product of just our brain requires that it goes to the grave when our physical form fails, and we die.

However, there is a tremendous amount of data that contradicts these beliefs, and this data has become so voluminous that not considering the possibility of something more is becoming the equivalent of burying one's head in the sand. It is important to understand that this perspective is now more a choice than something supported by data and facts.

As a scientist, I am here to tell you that the world of mind and spirit is real. It is as real as your body and the rest of nature. We have been living in a paradigm that is too small for the modern world and this is causing society much harm. According to the old materialistic form of thinking we are separate from everyone else and from everything around us. This road leads to depression and destruction. The paradigm developed in this book reimagines who we are and proposes that we are all connected to each other and all of nature. We all play a meaningful role within a much grander universal play. Death is no longer the end but a doorway to something more. This road leads to understanding, empowerment, hope and love.

In the vast web of Universal Consciousness, we encounter the principle of synergy and emergent properties. Although the concept of consciousness being a universal property may seem challenging to some of you, we can take a step further here and say that the whole is greater than the parts.

In other words, beyond the bits of consciousness you and I contribute to the whole and the subtle form of consciousness found throughout the universe, it is possible these all come together synergistically into One (the singularity), and as one conscious state form a Being with an existence, thoughts, and intensions that go beyond anything we can imagine.

A powerful story from the Jewish Torah sheds light on this concept of Universal Consciousness. In Exodus, Moses encounters God on sacred ground, and when asked about God's name, the response is the ancient Hebrew phrase "YHWH," meaning "I AM." This name represents an eternal, transcendent being—a Universal Consciousness that permeates all of existence. We are all intimately connected to this great I AM, we are known and valued entirely by Universal Consciousness. Love lies at the heart of this connection, uniting us in a profound and meaningful way. Initially, you may be tempted to devalue this story as a myth, but myths often point us to deep truths about the universe.

As we grasp the concept of I AM as the ultimate point of connection for all consciousness, we begin to understand the importance of PSI phenomena—the experiences that go beyond our individual perceptions.

Our ability to connect with others through PSI is completely dependent on our connection to Universal Consciousness. By prioritizing this intimate link with I AM, we foster a deeper bond with each other, transcending prejudices and embracing our shared humanity.

The connection we share through I AM is emphasized by stories of reincarnation. These stories allow us to see the intimate connections we share with each other. To the point that we feel that we are each other. Let that sink in. It gives a deeper meaning to the saying, "Do onto others as you would have them do onto you." Also, it may help you reconsider all the ways we separate each other. How can we hold prejudices when we are intimately connected to the people we may reject?

Amidst the wisdom of these revelations, one powerful message echoes throughout: I AM is synonymous with love. The story of Yeshua exemplifies this truth, as I AM sent his son to live among us, bringing us closer and emphasizing the importance of love in our lives. John, Chapter 3 from the Bible tells us:

[16] For God so loved the world that he gave his one and only Son, that whoever believes in him shall not perish but have eternal life.[17] For God did not send his Son into the world to condemn the world, but to save the world through him.

This is how much I AM loves us. This is how much Universal Consciousness wants us to stay connected. Not for the benefit of I AM but for ours. Step back for a moment and think about this. Universal Consciousness, the great I AM, the ultimate source and point of connection of all that is, was willing to enter our fractured world and live with us. To live as a normal human being, share this message of connection and eventually die for our salvation, that we might know him more dearly and gain a much deeper understanding of his love for us.

The story here is transformative. By engaging with this framework, we can clearly see the idea that we are alone and separate is not only incomplete but is completely wrong. It is important for us to see that our separateness is a constant message we receive from the dominant culture. If we focus on what we can touch, taste, see, hear, and smell in the traditional sense, we are locked into a system of understanding that devalues our connection to each other and the great I AM.

This system of focusing on the material world, our bodies, wealth, and power is what is making us so sad. We have disconnected ourselves from the source of all that is important, and we find that our lives have little meaning. We feel isolated and alone.

We have so many in the world that are working to create peace by so many different means but none of it works on a large scale or for very long. Many cultures have tried to create peaceful societies through strength (Pax Romana or Pax Britannica), but this is always unstable. Some societies have tried to create peace through philosophical understandings (Buddhism, Taoism) but something is still missing. The only way to create true, stable peace is by creating a society that is based on a fundamental understanding of I AM, our connections to I AM and through I AM our connections to each other.

Some religious philosophies give the impression that I AM is an impersonal force or disincarnate essence that pervades the universe. This perspective misses one of the most important aspects of a Universal Consciousness who is a being of unimaginable presence, love, and grace that desires to have a personal relationship with each and every one of us. This relationship, like any relationship, comes with expectations. Principal among these expectations is that we will work on the issues, hurdles and blockages that keep us from having a deeper more connected relationship with I AM. There is an expectation of transformation.

All the great challenges we face in the world arise from our inability to clearly see that we are not separate but are intimately connected through I AM. By understanding that we are part of a large whole, we gain an intuition that what I do in the world significantly affects us all.

The way I eat, the way I live, the things I value all affect the entire world and it matters. The emissions coming out of my vehicle affect us all, and as we currently see this is changing our climate. The trash I throw away may end up in the ocean and this is creating massive pollution islands. The products I buy may be produced in a fashion that pollutes the environment and this is likely harmful to the local population where they are produced.

I don't say these things to make you feel bad, but to show you the natural consequence of not understanding that we are all connected and part of Universal Consciousness.

So, let us embrace the great I AM and the source that unites us all. Let love guide our steps, empower us to see beyond ourselves, and kindle everlasting peace within and throughout the world. Together, we embark on a transformative journey—one that transcends boundaries and leads us to the heart of our interconnected existence.

I leave you with a few words from the Bible. The following is a quote from the Gospel of John from the Bible. It can be found in Chapter 17, verses 20-23 and they record some of the last words spoken by Yeshua as he prays for us all:

[20]"My prayer is not for them alone. I pray also for those who will believe in me through their message, [21] that all of them may be one, Father, just as you are in me and I am in you. May they also be in us so that the world may believe that you have sent me.[22] I have given them the glory that you gave me, that they may be one as we are one—[23] I in them and you in me—so that they may be brought to complete unity. Then the world will know that you sent me and have loved them even as you have loved me."

ABOUT THE AUTHOR

Dr. Alex Escobar has been developing the Quantized Visual Awareness model as a hypothesis for visual consciousness. He has published several articles on his research. If you are interested, please refer to the following:

Escobar, WA., Slemons, Megan (2020). Could Striate Cortex Microcolumns Serve as the Neural Correlates of Visual Consciousness?, *Athens Journal of Sciences,* 7, 127-142.

Escobar, A. (2016). QVA: A Massively Parallel Model for Vision. *Psychology of Consciousness: Theory, Research, and Practice,* 3(3), Sep 2016, 222-238

Escobar, W.A. (2013). Quantized Visual Awareness. *Frontiers in Psychology ,* 4, 1-11.

Escobar, A. (2011). Qualia as the Fundamental Nature of Visual Awareness. *Journal of Theoretical Biology, 279,* 172-176.